Manual of Water Supply Practices

# M28

# Rehabilitation of Water Mains



Third Edition

American Water Works Association

Manual of Water Supply Practices — M28, Third Edition

# Rehabilitation of Water Mains



Disclaimer
The authors, contributors, editors, and publisher do not assume responsibility for the validity of the content or any consequences of its use. In no event will AWWA be liable for direct, indirect, special, incidental, or consequential damages arising out of the use of information presented in this book. In particular, AWWA will not be responsible for any costs, including, but not limited to, those incurred as a result of lost revenue. In no event shall AWWA's liability exceed the amount paid for the purchase of this book.

AWWA Senior Manager of Editorial Development and Production: Gay Porter De Nileon
AWWA Senior Technical Editor/Project Manager: Melissa Valentine
Cover Art: Cheryl Armstrong
Production: Janice Benight Design Studio
AWWA Senior Manuals Specialist: Molly Beach

If you find errors in this manual, please email books@awwa.org. Possible errata will be posted at www.awwa.org/resources-tools/resource.development.groups/manuals-program.aspx.

**Library of Congress Cataloging-in-Publication Data**

Rehabilitation of water mains. -- Third edition / [edited by] Jon Turner, Mike Queen, Leonard Assard.
pages cm. -- (AWWA manual ; M28) (Manual of water supply practices ; M28)
Includes bibliographical references and index.
ISBN 978-1-58321-970-6 (alk. paper)
1. Water-pipes--Maintenance and repair. 2. Water-pipes--Cleaning. 3. Water-pipes--Linings. I. Turner, Jon (Hydraulic engineer) II. Queen, Mike. III. Assard, Leonard.
TD491.R395 2014
628.1'50288--dc23
2013042949

Printed in the United States of America

ISBN-13 978-1-58321-970-6 eISBN-13 978-1-61300-248-3



American Water Works Association

American Water Works Association
6666 West Quincy Avenue
Denver, CO 80235-3098
awwa.org

# Contents

# Figures

# Tables

This page intentionally blank.

# Foreword

The water industry in the United States is faced with quite a challenge. Much of its buried infrastructure is aged and approaching the end of its useful life. Water system pipeline infrastructure primarily installed in the 1950s and 1960s or earlier has served its communities well. Across the country, water mains provide water to communities with day and night reliability except for the occasional water main interruption. However, that is changing as the pipelines have met or are already exceeding the expected life spans of the pipeline materials. Increasing failures of aged pipeline networks may reach a performance level unacceptable to customers and burdensome to the operators of the systems.

Conventional pipeline replacements of older pipeline infrastructure have not occurred fast enough, largely owing to financial constraints. Failures are now impacting major transmission mains. As a result, communities now are faced with an almost insurmountable challenge to perform the inevitable task of replacing these key and critical water main pipelines. Coupled with reductions in revenue and a slowly recovering economy, the task is especially daunting. Water main rehabilitation may offer part of the solution by renewing more pipe at less cost.

Water main rehabilitation is not a new concept. Cement–mortar lining of water main pipelines is one of the oldest rehabilitation methods. The oil and gas industry was the first to develop systems and techniques beyond cement–mortar lining to rehabilitate critical pipeline networks without the use of conventional open-trench construction. The wastewater industry was the next utility group to firmly adopt pipeline rehabilitation techniques. The expansion of various technologies among these pipe-focused industries created numerous companies, processes, and materials as many groups scrambled to "build a better mousetrap." The fervor that resulted saw the installation of millions of feet of pipeline rehabilitation products in communities across the country and overseas.

The water industry, particularly in the United States, is the last to fully realize the economic, social, and community benefits of rehabilitation. Conventional construction has its place and continues to be utilized on a widespread scale; however, water main and water service rehabilitation has been steadily increasing in use and effectiveness. Companies that were engaged in the wastewater pipe rehabilitation process expanded their technologies to develop products that could withstand the internal pipeline pressures as well as be approved for potable water use. The number of rehabilitation projects being performed is expected to grow as the price of raw materials has risen and competition for limited space in public right-of-ways has increased with other buried utilities (including systems that have moved from aerial to burial).

The second edition of the M28 *Water Main Rehabilitation* manual, published in 2001, provided an expanded exposure of the technologies present at the time. The intent of the manual was to provide an overview of the processes used, considerations for what rehabilitation process to use, and suggestions for a successful project. This third edition of the manual continues with the same intent. It updates technologies since the last edition, expands categories to encompass available, proven processes, and provides exposure to other rehabilitation considerations such as keyhole technology and cathodic protection. The goal of the manual is to provide engineers, contractors, and decision makers with an overview of the processes used for water main rehabilitation. It is recognized that different pipe installations and types will warrant different rehabilitation options. These considerations are explained in detail including a decision flow chart to aid in the process and get the user to a narrow, more situation-specific set of options for their project.

It is recognized that the water main rehabilitation industry is continually offering new technologies to assist communities with rehabilitating their water pipeline infrastructure. This manual specifically addresses only those technologies with a proven track record within the water industry that can be employed by a water utility to successfully rehabilitate water mains. No attempt was made to evaluate one method of water main rehabilitation over another, and no attempt was made to evaluate relative costs between competing systems. It remains for the user of the manual to decide which system will best suit a specific project. In 2011, the AWWA Water Main Rehabilitation Committee decided to not consider any more new and upcoming technologies for chapter discussion. Because the field is up and coming, there are always new technologies that have been tried overseas that are taking hold here in the North American market.

The chair of the committee, Jon Turner, would like to thank the Executive Committee in place at the time of the manual preparation and to thank contributing authors responsible for new material supplied to continue to provide an up-to-date water main rehabilitation manual for water industry professionals.

## Executive Committee

Jon Turner, PE, Chair
Mike Queen, Vice Chair
Leonard Assard, Secretary
George Mallakis, Past Chair

# Acknowledgments


The Water Main Rehabilitation Committee acknowledges these individuals for their persistence and dedication as standing subcommittee members assigned to take the lead for updating the M28 manual:

### Executive Committee

*Jon Turner, PE, Chair,* Phoenix Civil Engineering, Inc., Ventura, Calif.
*Mike Queen, Vice Chair,* Consolidated Mutual Water Company, Lakewood, Colo.
*Leonard Assard, Secretary,* Heitkamp, Inc., Watertown, Conn.
*George Mallakis, Past Chair,* TT Technologies, Inc., Simi Valley, Calif.

*George Bontus,* PENG, Insituform Technologies, Edmonton, AB., Canada
*Richard "Bo" Botteicher,* Underground Solutions, Inc., Denver, Colo.
*Brien Clark, PE,* HDR/Schiff, Claremont, Calif.
*Dan Cohen,* Western Slope Utilities, Inc., Breckenridge, Colo.
*Steve Cooper,* Uni-Bell PVC Pipe Assocation, Louisville, Ky.
*Michael Davison,* Sanexen, Varennes, Que., Canada
*Benedict Ebner,* Heitkamp Inc., Watertown, Conn.
*Dan Ellison, PE,* HDR Engineering, Inc., Ventura, Calif.
*Dawn Flancher, PE,* Staff Engineer Liaison, AWWA, Denver, Colo.
*Steve Fox,* HDR/Schiff, Claremont, Calif.
*Michael Grahek, PE,* Los Angeles Department of Water and Power, Los Angeles, Calif.
*Behnam Hashemi,* IUPUI, Indianapolis, Ind.
*Kathie Hirata,* Los Angeles Department of Water and Power, Los Angeles, Calif.
*John Hohider,* Heitkamp Inc., Watertown, Conn.
*Derrick Horsman,* Alltech Solutions Inc., Moncton, N.B., Canada
*David Hughes, PE,* American Water, Voorhees, N.J.
*Joe Loiacono,* Sanexen Varennes, Que., Canada
*Tom Murli,* Underground Solutions Inc., Poway, Calif.
*Kenneth Morgan, PE,* Town of Gilbert, Gilbert, Ariz.
*Keith Oxner, PE,* Inliner Technologies, LLC, Chesterfield, Mo.
*Tom Rockaway,* University of Louisville, Louisville, Ky.
*Brian Rohan, PE,* Rohan Engineering, PC, Merrick, N.Y.
*Ryan Rogers,* 3M Company, Saint Paul, Minn.
*Jeremy Ross, PE,* Denver Water, Denver, Colo.
*Laurance Vaisey,* Cempipe Ltd., New Haven, England, United Kingdom
*Marc Wegner, PE,* HDR Engineering, Inc., Claremont, Calif.

This page intentionally blank.

# Chapter 1

# Pipeline Renewal Methods

Pipeline renewal is typically accomplished by one of two approaches: rehabilitation or open-trench construction, although other trenchless methods are also used. Trenchless technology is a type of subsurface construction work that requires little or no surface excavation and no continuous trenches. This chapter provides guidance in selecting between rehabilitation and open-trench construction and in determining which rehabilitation method is most appropriate for meeting goals.

The renewal of water mains is performed for three primary reasons:

1. Water Quality Improvement: to improve the quality of the water received by the consumer
2. Hydraulic Improvement: to increase the hydraulic capacity of the pipeline
3. Structural Improvement: to reduce leakage, decrease repair frequencies, lessen risk of property damage, and improve reliability.

Compared with conventional open-trench replacement, pipeline rehabilitation methods are often less expensive and less disruptive to the community; however, rehabilitation is not appropriate for all situations.

As described in other chapters in this manual, many different water main rehabilitation techniques exist, offering a variety of benefits. The best choice of method for each situation will depend on several factors, including: (1) the reason for the rehabilitation, (2) comparative costs, (3) site conditions, and (4) expected life-cycle performance.

This chapter provides guidance in selecting a pipeline renewal method, including a series of decision trees that can be used to help determine which types of methods should be considered.

## DISTRIBUTION SYSTEM WATER QUALITY IMPROVEMENT

Water main rehabilitation is frequently performed to mitigate water quality deterioration that occurs within the distribution system. The goal is to improve water quality at the point of use. The improvements can be very dramatic, particularly when the existing main is unlined cast iron.

In most cases, the water quality benefits achieved by the various pipeline rehabilitation methods are fairly equal. This assumes that the materials being employed are certified in accordance with NSF/ANSI Standard 61, and industry accepted standards are employed. It also assumes that the water is not particularly soft. When the water is soft, problems with high pH can occur if cement–mortar lining is used.

The quality of treated drinking water can vary considerably, both from system to system and within a system, as a result of deterioration after it leaves the treatment plant and comes in contact with the interior of distribution system piping. Over time, changes in water chemistry can cause problems throughout the distribution system, ultimately affecting the quality of the water delivered to the end user. The deterioration of water quality that occurs within the water distribution system is often signaled by customer complaints regarding the clarity, color, taste, and odor of the water. Although the water may be safe to drink, it may be aesthetically unpleasant due to sediment that has been stirred up or by biological processes that can thrive in highly scaled pipelines and impart an odor to the water. However, the concern is not always simply about aesthetics. Water stagnation and chlorine depletion occur in highly scaled pipelines, resulting in greater risk of coliform regrowth. The majority of these problems fall into three categories: sedimentation, scaling, and biofilm formation.

### Sedimentation

Sedimentation is the process whereby solids settle out of water moving at low velocity in a main, reducing interior cross section and capacity and becoming a potential source of customer complaints about water quality. Source water pipelines or pipelines carrying unfiltered or improperly treated water can be subject to deposits of sand, silt, or organic materials. In pipelines receiving well water, particulates from oxidation of iron or manganese are also common, if the water is not filtered. In extreme cases, sedimentation can also contribute to hydraulic problems, particularly at low points in the pipe. The most common source of sediment is the internal corrosion of the pipeline itself.

In smooth-walled pipelines, sediment generally moves through the system at moderate flow velocities and does not accumulate. However, where the pipeline is heavily scaled, sediment settles into the recesses of the scale and builds up over time. This sediment can then be stirred up when the flow velocity increases (e.g., a fire hydrant is opened) or the direction of flow reverses. The result can be severely discolored water as a large volume of sediment becomes suspended and is delivered to the customer's tap.

### Scaling or Tuberculation

Scaling is the formation of hard deposits on the inside wall of the pipe. These deposits are frequently the by-product of pipe corrosion, wherein iron combines with calcium and other minerals within the water to form *tubercles*. The process, often called *tuberculation*, is assisted by bacteria within the scale that feed on iron leached from the pipe. Although scaling is most pronounced in cast-iron pipes, it is also commonly found in unlined steel pipes, copper pipes, concrete, and asbestos–cement pipes. Figure 1-1 shows a fairly typical tuberculated cast-iron pipeline.

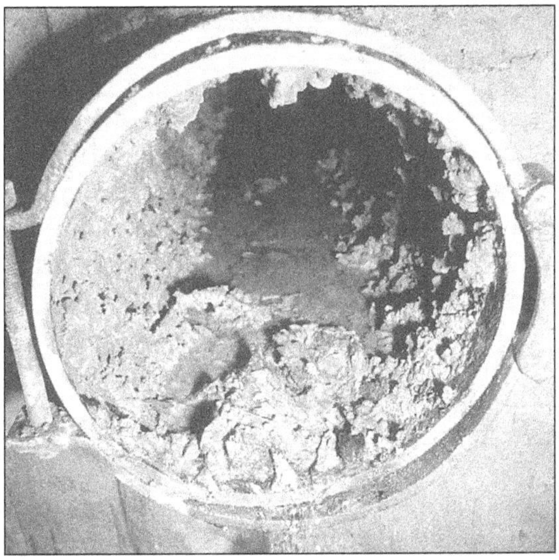

Figure 1-1 Pipe with tuberculation caused by corrosion

Before the 1960s, many iron pipes were installed without effective, long-lasting linings to protect the interior surfaces. These pipes often experience internal corrosion and develop tubercular scales. This scaling restricts the flow and creates areas where sediment is deposited, and chlorine depletion occurs.

Water discoloration complaints can occur when these sediment deposits are stirred up as previously described; however, discoloration can also occur if the corrosion activity within the pipeline is particularly high. If the scales are removed and a lining is not subsequently installed, exposure of the underlying iron or steel pipeline often results in increased corrosion activity and more frequent complaints about water discoloration. Such corrosion activity can be controlled to some extent through water chemistry (corrosion inhibitors), but cleaning of pipelines without lining is generally not recommended.

### Biofilm Formation

Biofilms can develop within pipe made from any type of material; however, biofilms are most common within highly scaled cast-iron pipe, where sediments and recesses allow iron-reducing bacteria to thrive in the absence of effective disinfectant. As the pipelines corrode and tubercles develop, the hospitality of the environment for biofilm increases. The greater the roughness of the pipe surface, the harder it is for an effective disinfectant residual to be maintained near the pipe surface. The reduction of the iron leached from the pipe also provides the energy source for the bacteria.

Biofilms also form readily in raw water systems or portions of the finished water system where water is high in iron or manganese or other nutrients. Such biofilms take the form of slimes, i.e., soft and filamentous. Even where scales do not form, these biofilms can severely affect water clarity and produce taste-and-odor problems. They can also significantly diminish hydraulic capacity.

## HYDRAULIC IMPROVEMENT

Increasing roughness and the buildup of scale or slime inside water distribution piping can greatly reduce the hydraulic performance of the system. This can significantly impact the ability to deliver adequate fire flows and can also affect pressures and flows available to customers.

Hydraulic engineers are able to calculate head losses and flow in pipes using the empirically derived Hazen–Williams formula, which relates flow to the physical properties of the pipe and pressure changes due to friction; however, the Hazen–Williams equation cannot be applied to all fluids in all conditions. It is only valid for ambient temperatures (40°F to 75°F [4.4°C to 23.8°C]) and at turbulent flow (Reynolds numbers above $10^5$). For liquids outside these parameters, the Darcy–Weisbach formula is more reliable for frictional head loss calculations at steady-state flow. In more complicated instances, computer models based on Hardy Cross are more accurate. For the discussion in this manual, the Hazen–Williams equation is:

$$V = kCR^{0.63}\, S^{0.54} \qquad \text{(Eq. 1-1)}$$

where:

$V$ = velocity, ft/sec (m/sec)
$k$ = conversion factor
$R$ = hydraulic radius, ft (m), which is the cross-sectional area of the pipe divided by the wetted perimeter
$C$ = Hazen–Williams roughness coefficient
$S$ = slope of the hydraulic grade line, ft/ft (m/m)

*C* is a measure of the roughness of the interior of the pipe. Expressed in terms of *C*, the formula can be stated several ways. Once such way is stated as:

$$C = 2{,}466QD^{-2.63}\, H^{0.54}\, L^{0.54} \qquad \text{(Eq. 1-2)}$$

where:

$C$ = Hazen–Williams roughness coefficient
$Q$ = quantity of flow in a pressure conduit, mgd ($m^3/d$)
$D$ = nominal diameter of the pipe, in. (mm)
$H$ = head loss, ft (m) of water
$L$ = length of pipe, ft (m)

The Hazen–Williams *C* factor, and hence the flow in a pipeline, depends on the type of pipe and its interior condition (see Table 1-1). For a given velocity, increased internal surface roughness (changing laminar to turbulent flow) leads to greater pressure loss. By measuring pipe flows and pressure changes between two points along a pipeline, operators can calculate Hazen–Williams *C* factors and determine the degree the pipeline has become roughened and constricted. These data help in making informed decisions about which process to employ to restore hydraulic efficiency. Collecting data for the Hazen–Williams *C* factor after employing a cleaning or pipe rehabilitation process is also a useful way to gauge the impact of the system improvements.

**Table 1-1 Hazen–Williams roughness coefficient**

| Condition | *C* |
|---|---|
| New pipe | 130–140 |
| Fair to normal (interior clean) | 100 |
| Significant reduction in pipe capacity | 70 |
| Severe problem—interior cross section greatly reduced | 30–50 |

Pipeline rehabilitation very frequently results in significant improvement of system hydraulics, particularly where a cast-iron main is choked with tuberculation. Not only is a smoother pipe surface achieved, but at times, the effective flow cross section can be increased significantly.

Because the various methods use different materials and result in different internal diameters, the various methods achieve different degrees of hydraulic improvement. Table 1-2 provides a general comparison of the hydraulic improvements that the various rehabilitation methods can provide.

Moderate hydraulic improvement can also be achieved by some of the pipeline cleaning methods.

## STRUCTURAL IMPROVEMENT

The structural performance of water mains deteriorates over time due to several causes. Cast-iron, ductile-iron, and steel piping are subject to internal and external corrosion, resulting in pitting and wall thinning, which can lead to leaks and eventual burst failures. Cement-based pipes such as asbestos–cement and concrete pipe may also be subject to deterioration due to corrosion of the cement matrix and/or steel reinforcement. In addition, all types of pipe, including plastic, may be subject to joint failure between pipe lengths and hence excessive leakage, which can in turn lead to washout of bedding and subsequent structural failure.

Such structural and leakage failures can have direct consequences such as high repair costs, water quality problems, service interruptions, and loss of treated water. They may also have indirect consequences in terms of the economic damage associated with pipe bursts and the public relations damage to the service provider.

The structural improvements afforded by the techniques discussed in this manual vary considerably. Cement–mortar lining and epoxy lining are generally considered *nonstructural* because they offer very minor structural improvements at best. Other methods arguably offer the same structural integrity achieved by a new pipeline installed using conventional open-trench construction. In selecting a pipeline rehabilitation method, one of the key considerations is matching the method to the pipeline. A nonstructural method (i.e., Class I*) is absolutely appropriate for a pipeline that has experienced very little deterioration, but this method would not be appropriate where external corrosion has caused significant weakening of the pipe and where this corrosion is expected to continue.

**Table 1-2 General comparison of hydraulic improvements**

| Hydraulic Improvement Anticipated | Rehabilitation Method |
|---|---|
| Modest hydraulic improvement | Loose sliplining |
| Moderate hydraulic improvement | Cement–mortar lining<br>Epoxy and other polymer lining<br>Modified sliplining (close-fit)<br>Cured-in-place lining |
| Greater hydraulic improvement | Pipe bursting |
| Unlimited hydraulic improvement | Open-trench construction |

* See appendix A for definition of Class I, II, III, and IV linings.

## WATER MAIN CONDITION EVALUATION

Before employing a nonstructural or semistructural method, an evaluation of the structural condition of the water main is warranted. This evaluation can range from simple and inexpensive, to high-tech and quite costly. The more valuable the pipeline, the more time and money the utility should invest in making the right decision.

The following methods have been used successfully to guide decisions regarding pipeline renewal:

- **Leak/break performance.** Where repair records indicate that a pipeline has had few or no corrosion-caused failures and the pipeline has been in service for many decades, it is often assumed that external corrosion activity is minimal, and the pipeline is a good candidate for a nonstructural lining.
- **Sample extraction/evaluation.** In the United Kingdom, pipeline samples are exhumed then grit blasted to remove graphitization and expose the corrosion pits. The remaining life of the pipeline is then estimated using a method that examines pit depth and pit spacing. One U.K. utility replaces the pipe, for instance, if the estimated remaining life is less than 20 years, and uses a nonstructural lining where the life expectancy is 30 years and more. Semistructural linings are used for those in the middle, with life expectancies between 20 and 30 years.
- **In-situ testing.** For pipelines where a greater investment is warranted, nondestructive evaluation methods should be considered. Depending on the type of pipe, remote field technology or remote field eddy current can be used to find areas of weaknesses throughout the pipeline. Other techniques have been used for spot evaluations at locations of particular concern. Acoustic detection methods of various types are used to search for leaks or detect the sounds of incipient failure.

## PRIORITIZATION

Because budgets are always limited, a method of prioritizing work is important. If a primary driver for the renewal program is structural improvement, the pipelines that pose the highest risk should receive the highest priority. In assessing risk, it is helpful to recognize that risk has two components: probability and consequence. For something to be risky, it must be both likely to occur and have significant consequences. This concept is often expressed mathematically:

$$\text{Risk} = \text{Probability} \times \text{Consequence} \qquad \text{(Eq. 1-3)}$$

To perform this mathematical calculation requires more data (and better data) than are typically available. So instead, it is often helpful to look at *relative* risk as expressed in Table 1-3.

Prioritization can also be performed using regression analysis. With sufficient data regarding the age, soil conditions, pressure, pipe characteristics, etc., effective statistical models have been developed for specific systems. However, regression models built for one system have not been demonstrated to work on other systems.

Table 1-3 Relative risk assessment

| Probability | | Consequence: Low | Consequence: Medium | Consequence: High |
|---|---|---|---|---|
| | High | Repair on failure | Schedule renewal | Fix now |
| | Medium | Repair on failure | Assess proactively | Schedule renewal |
| | Low | Repair on failure | Monitor | Assess proactively |

## COSTS AND BENEFITS

Many factors influence the cost of a water main renewal project. Some of the factors are: project size, pipeline size, method used, bypass system requirements, traffic conditions, number of laterals, number of valves or fittings, paving requirements, etc. Costs are also influenced by the availability of local contractors who have the equipment and knowledge needed to perform the rehabilitation. Generally, the less-structural spray-applied methods (Class I) will be less expensive than more fully structural (Class IV) methods.

The cost of a rehabilitated pipeline typically ranges from 25 percent to 100 percent of the cost of conventional open-trench construction. However, even where there are no significant cost savings, rehabilitation may still be preferred because it results in fewer construction impacts to the community.

When properly applied to an appropriate pipe, the life expectancy goal of a rehabilitated pipeline should be similar to that of a new pipeline—50 to 100 years.

## REHABILITATION SOLUTIONS

This manual describes several possible solutions to problems arising from corrosion and deposition of internal scales. These range from simple periodic cleaning to replacement of the pipe using trenchless techniques. All of the solutions discussed in the manual make some use of the existing pipe, either as part of the rehabilitated system (renovation solutions) or as a convenient route for installation of new piping (replacement solutions). Solutions involving installation of a replacement pipe along a new route, such as open-trench laying, directional drilling, and microtunneling, are outside the scope of this manual.

Selecting the optimal solution to a specific pipeline problem is a complex process involving both technical and economic considerations. Both the Water Research Foundation and several AWWA Technical and Educational Council (TEC) committees are developing computer-based decision tools to assist utility engineers in this process. This work is expected to come to fruition while this edition of the manual remains in effect. In the meantime, the following guidelines may prove useful.

## SELECTION OF REHABILITATION SOLUTIONS

Key elements in the selection of a rehabilitation solution are:

1. The exact nature of the problem(s) to be solved
2. The hydraulic and operating pressure requirements for the rehabilitated main
3. The materials, dimensions, and geometry of the water main
4. The types and locations of valves, fittings, and service connections
5. The length of time in which the main can be taken out of service
6. Site-specific factors

The aim of the selection process is to consider all these factors and to arrive at the most cost-effective, technically viable solution. Ideally, the cost estimate should include not only direct contracting and related costs but also indirect costs associated with public disruption and longer-term maintenance and other life-cycle costs.

One approach to rehabilitation method selection is summarized in Figures 1-2, 1-3, and 1-4. Together, these charts provide a framework for selecting or rejecting groups of rehabilitation methods, depending on the nature of the performance problems, hydraulic requirements, and some site-specific factors. In some cases, the charts indicate use of lining techniques classified as either Class I (nonstructural), Class II/III (semistructural), or Class IV (structural). A more detailed discussion of this classification system and of other key design issues associated with such lining techniques is presented in appendix A.

The figures do not list cleaning as a solution for water quality or flow and pressure problems. Cleaning with one of the various techniques discussed in the manual may well offer the lowest-cost immediate solution to many of these problems. It may offer a long-term solution if repeated as required or combined with chemical treatment of water to prevent or delay recurrence of the original problem. However, cleaning is more frequently used as a necessary preliminary step before carrying out one of the lining processes described in the manual.

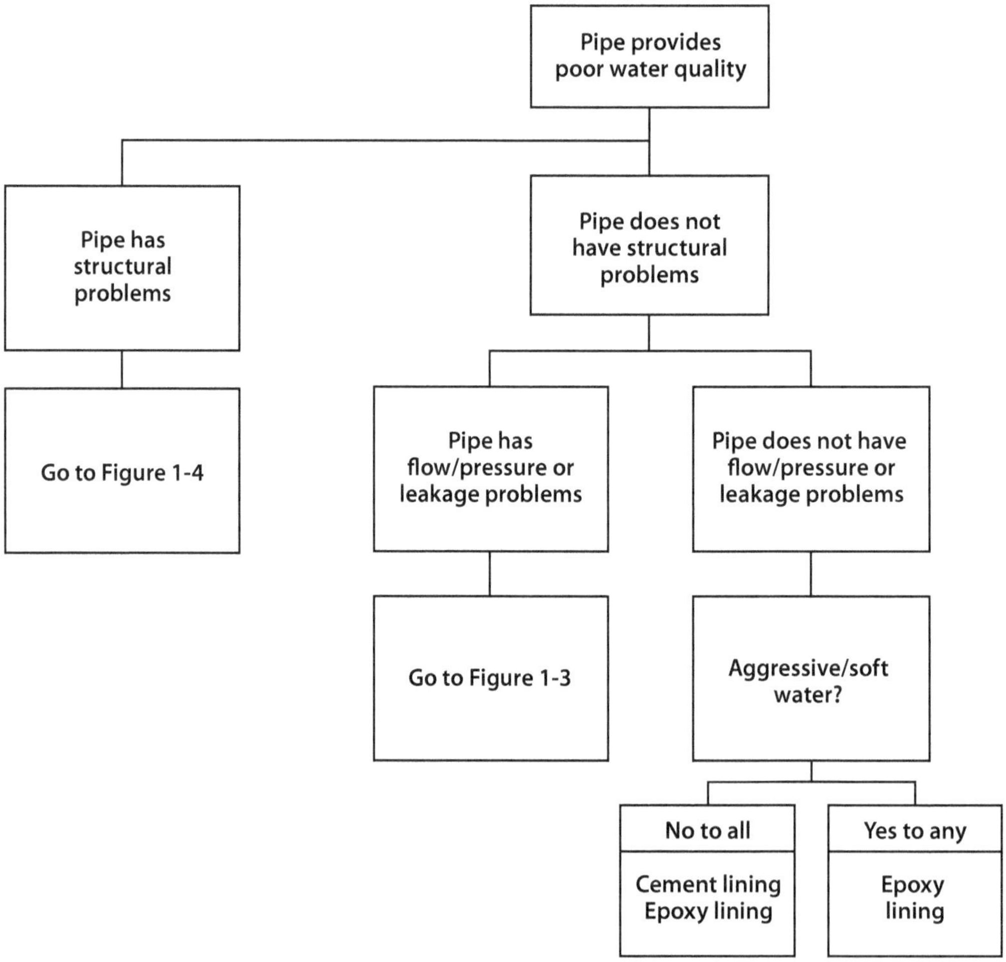


Figure 1-2 Selection of rehabilitation techniques to resolve water quality problems

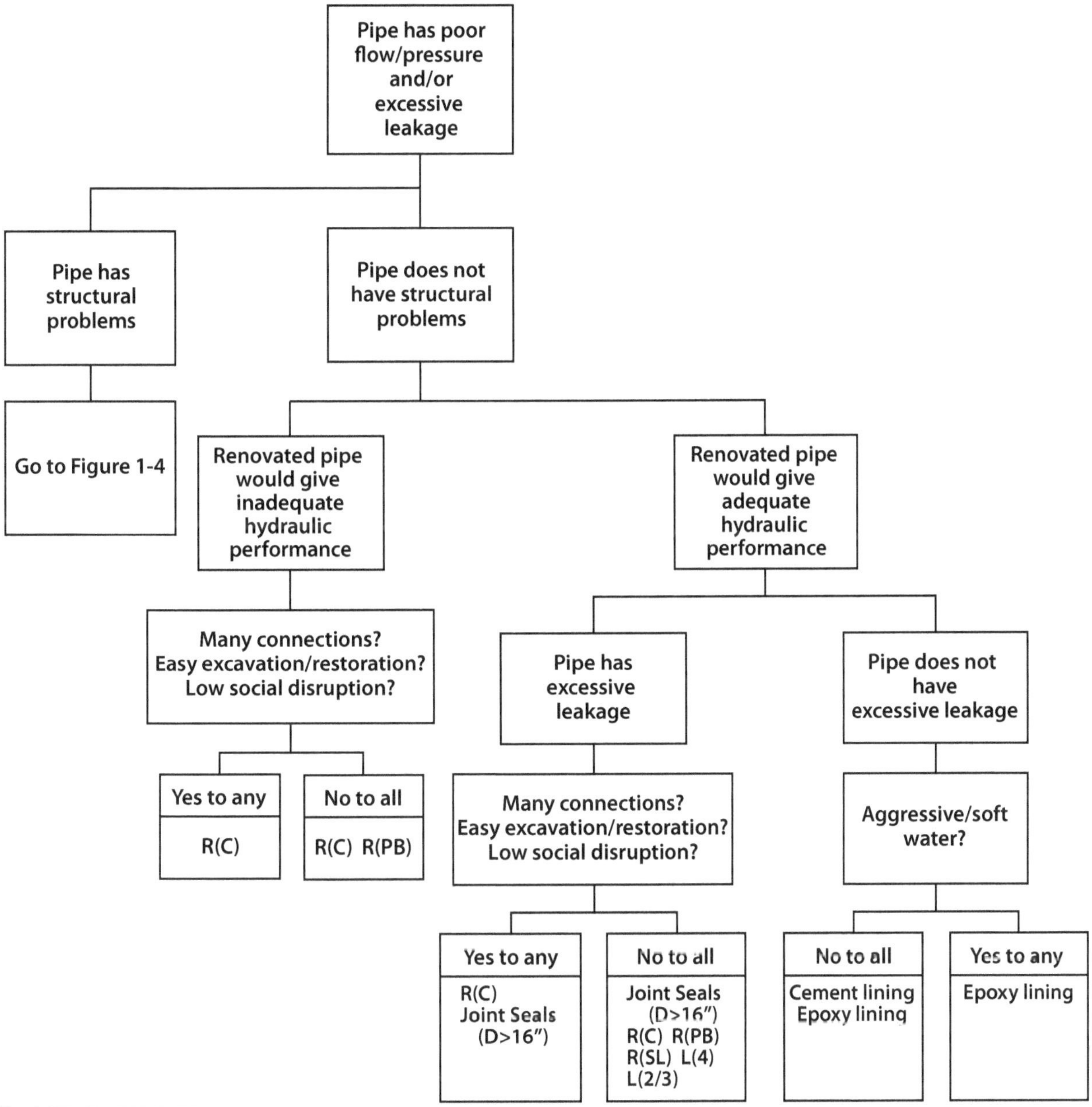


Notes:

R(C)—Replacement (conventional or boring/directional drilling)
R(PB)—Replacement (pipe bursting)
R(SL)—Replacement (sliplining)
L(2/3)—Lining (semistructural—Class II/III)
L(4)—Lining (structural—Class IV)

Figure 1-3 Selection of rehabilitation techniques to resolve flow, pressure, and leakage problems

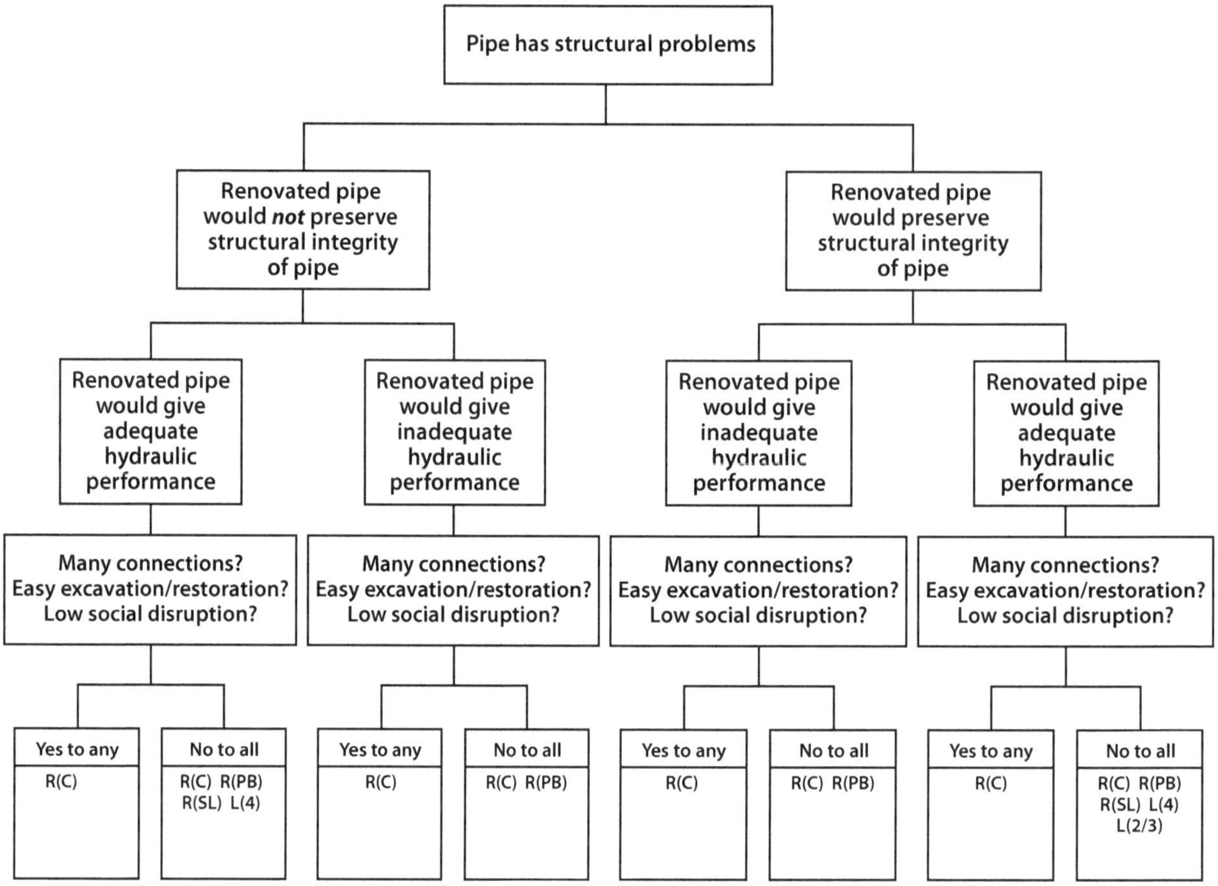


Notes:
- R(C)—Replacement (conventional or boring/directional drilling)
- R(PB)—Replacement (pipe bursting)
- R(SL)—Replacement (sliplining)
- L(2/3)—Lining (semistructural—Class II/III)
- L(4)—Lining (structural—Class IV)

Figure 1-4 Selection of rehabilitation techniques to resolve structural problems

## REFERENCE

American National Standards Institute (ANSI). 2012. Drinking Water System Components—Health Effects. NSF/ANSI 61-2012. 198. Ann Arbor, Mich.: NSF International.

Chapter 2

# Preconstruction Activities

## ADVANCE PLANNING CONSIDERATIONS

Prior to undertaking a water main rehabilitation project or program, advance planning should be undertaken to fully understand the effects such a project will have on the affected customers, the community at large, and with other public agencies, and to determine strategies to mitigate the negative impacts such projects produce.

The reasons a utility decides to undertake water main rehabilitation will vary (see chapter 1 for discussion on water main rehabilitation techniques) and the method chosen will reflect those parameters. However, it is important to recognize that different water main rehabilitation techniques may impact the community differently. Therefore, it is important to have a clear understanding how each water main rehabilitation technique is performed and what the impacts will be to the community.

Once the water main rehabilitation technique has been determined, the utility should begin to develop a scope of work for the project and determine the overall project boundaries or hire an agent to do so. Once all the affected streets have been determined, the project team should begin the coordination process with other public agencies, such as the street paving department, to ensure that any streets that are planned for resurfacing can be rescheduled so the resurfacing work is done after the water main rehabilitation work has been completed. Likewise coordination with the other public agencies should be planned to prevent competing construction projects within the same right-of-way. The local governmental representative should also be notified of proposed work within their area of responsibility.

Because weather may play a key role in the success of certain water main rehabilitation techniques, it is important that projects are planned with seasonal weather patterns in mind. Projects that encompass sensitive environmental areas need additional planning and mitigation efforts to avoid negative impacts to those areas.

## PREPARATION OF PLANS AND SPECIFICATIONS

During the preparation of plans and specification for a water main rehabilitation project, a thorough research should be conducted of the utility's records to identify the locations of the water main, valve, services, fire hydrants, and other appurtenances that will be affected by the water main rehabilitation project. Failure to identify and locate the affected water mains and appurtenances can lead to significant project delays and costly change orders during the water main rehabilitation project. A substructure investigation is critically important to identify all other substructures located within the project boundaries, because unidentified substructures can have the same negative impact on the project schedule and costs as unidentified water mains and appurtenances.

Most work on water mains occurs in a roadway that carries vehicular traffic. A study of the impact the project will cause to the flow of traffic should be performed. Some cities have traffic departments that are responsible for determining what traffic mitigation efforts will be required for any construction work that takes place within the roadway. The project team should include the requirements for proper traffic delineation in its plans and specifications.

The project team should have detailed plans and specifications indicating the beginning and ending points for the rehabilitation process, as well as the method of pipe closure when the rehabilitation process is complete. The specifications should also detail the minimum acceptable requirements for the rehabilitation process along with the pertinent ASTM (American Society for Testing and Materials) standards for the materials used in the rehabilitation process.

Because there will most likely be interruptions in water service to the utility's customers, the utility should have a customer notification process in place, so that the customers can be given ample notice of water service interruptions. The utility should also provide the customers with prior written notice of the project detailing the scope of work and the reasons the work should be done. The notice to the customers should also provide them with a telephone number that can be accessed for information or to report a water outage or other project-related problems.

During the job planning process, the project team should determine what additional work should be performed on the water distribution system while the project is underway. Water main rehabilitation projects provide an opportune time to replace water meters, water valves, fire hydrants, and other ancillary water-works materials.

## WATER MAIN REHABILITATION CONTRACTS

Once the plans and specifications have been produced, and the necessary permits obtained from the various agencies involved, the project team will be ready to obtain a water main rehabilitation contractor for the project. Because there are many different requirements for different cities when it comes to bidding, it would be difficult to outline a process here, but there are several points that are common to every water main rehabilitation project.

The project team should endeavor to hire a qualified water main rehabilitation contractor with sufficient experience and expertise in selected water main rehabilitation process to ensure the project will be completed with satisfactory results. The project team may want to establish minimum experience levels for the contractor before being allowed to bid on the contract.

The project team should advertise the project in the appropriate trade magazines or other such venues to attract contractors with the necessary experience. Other utilities can be contacted to obtain contractor references. It is also advisable that the project team have its own quality control/quality assurance officer involved in the process.

This page intentionally blank.

Chapter 3

# Maintaining Service

The purpose of any water supply system is to maintain a continuous supply of safe water sufficient for customer's needs, including fire protection. This purpose remains of primary importance even as rehabilitation work is performed. As noted previously in this manual, depending on the cleaning or lining method used, temporary distribution system shutdowns may occur. Some processes require relatively brief shutdowns, so work can be completed without installing bypass lines. Conversely, some cleaning techniques and all lining methods require more extensive shutdowns that may create the need to install bypass piping. In cases where rapid return to service linings are employed, special disinfection practices and boil water advisories may be necessary if bypass piping is to be avoided.

## BYPASS PIPING

Installing bypass piping must be a carefully planned and well-coordinated procedure, and each project has unique considerations. The utility and contractor should jointly review the plan, with regard to customer service issues and traffic concerns. Other considerations include (1) individual service connections, (2) overall demand, and (3) fire protection demand (e.g., pipe sizing). Note that bypass lines are connected to fire hydrants or other temporary connections outside of the area of the shutdown.

The installation of bypass lines can be time consuming and labor intensive. However, the use of bypass piping does allow fairly long shutdowns while still maintaining acceptable service to customers. The hydraulic requirements of the portion of the water system to be removed from service will establish the parameters for sizing the bypass piping network. Temporary connections may be required for fire protection, depending on local jurisdiction.

After the piping has been sized and installed, it must be disinfected as directed in ANSI/AWWA Standard C651 before being placed in service. The piping and hoses used in bypass piping should be ANSI/NSF Standard 61 approved for use in potable water systems. Connections should always maintain positive pressure in the bypass pipes. Leakage should be minimized to limit damage, save money, and maintain customer confidence.

## Residential Installations

A residential area bypass line is usually 2 in. (50 mm) in diameter with provisions for a 0.75-in. (19-mm) or 1-in. (25-mm) hose connection to each residence along the main (Figure 3-1). The bypass hose is connected to the customer's service line though a meter pit, an outside hose bib on the dwelling, or interior pipes accessed through a basement window. The bypass line is usually installed along the curb or in the gutter so that it can be buried under driveway aprons and street intersections. Cold-mix asphalt, stone dust, or other suitable material can be used to cover and protect bypass piping in traffic areas. The exact location of the bypass line installation depends on the project-specific circumstances. Site conditions affect the amount of potential damage to the line, possible tripping hazards, and any obstructions to pedestrians and vehicular traffic.

## Commercial Installations

In a business or commercial area, bypass lines are usually 4 in. (100 mm) or larger in diameter. Additional precautions may be required to avoid problems with parallel parking of vehicles, damage to the bypass line, or damage by vehicle tires when parking or turning. Damage usually occurs because of in-line taps, couplings, or valves. Building connections are made through hoses similar to residential installations, except that larger hoses may be required. Depending on local street and traffic concerns, bypass piping may traverse driveways and street intersections by (1) burying temporary pipes in a shallow pavement trench, (2) placement on the street surface with asphalt or suitable material ramping, (3) boring under the street, or (4) running pipe through existing culverts, etc. Whenever possible, street crossings should be made perpendicular to the street.

## Reconnection of Service Lines

Care during the reconnection phase is very important to avoid service interruptions. Upon completing the cleaning or lining, the permanent pipelines must be flushed and disinfected according to ANSI/AWWA Standard C651. Following this step, the lines will be ready for service. The bypass lines can be disconnected and customers reconnected to the permanent service lines.

Some cleaning and lining operations may require that individual service lines be blown back to the main before reconnection; this can be accomplished by various means. High-pressure water can be forced back through the service line from inside the building to the main. Blowback can also use compressed air fed directly from a compressor or from portable worker-borne tanks. Note that all blowback air should be generated by oil-free compressors.

## Problems

Exposed bypass lines may be subject to overheating problems during periods of warm weather. In this situation, the bypass line may be coated with white or other heat-reflective paint to minimize solar heat gain, and/or provided with bleeder valves at the ends of the bypass to allow for periodic flushing.

Conversely, during periods of cold weather, freezing of temporary pipes may present problems. If scheduling permits, rehabilitation or cleaning projects should be conducted during more temperate months. However, if work during extreme cold weather is unavoidable, provisions must be made to maintain water flow in all bypass lines to prevent freezing.

The temporary network also represents a potential for emergency repairs due to hose ruptures or vandalism. As discussed in the next section, bypass service increases the need for good communication with various groups of consumers.

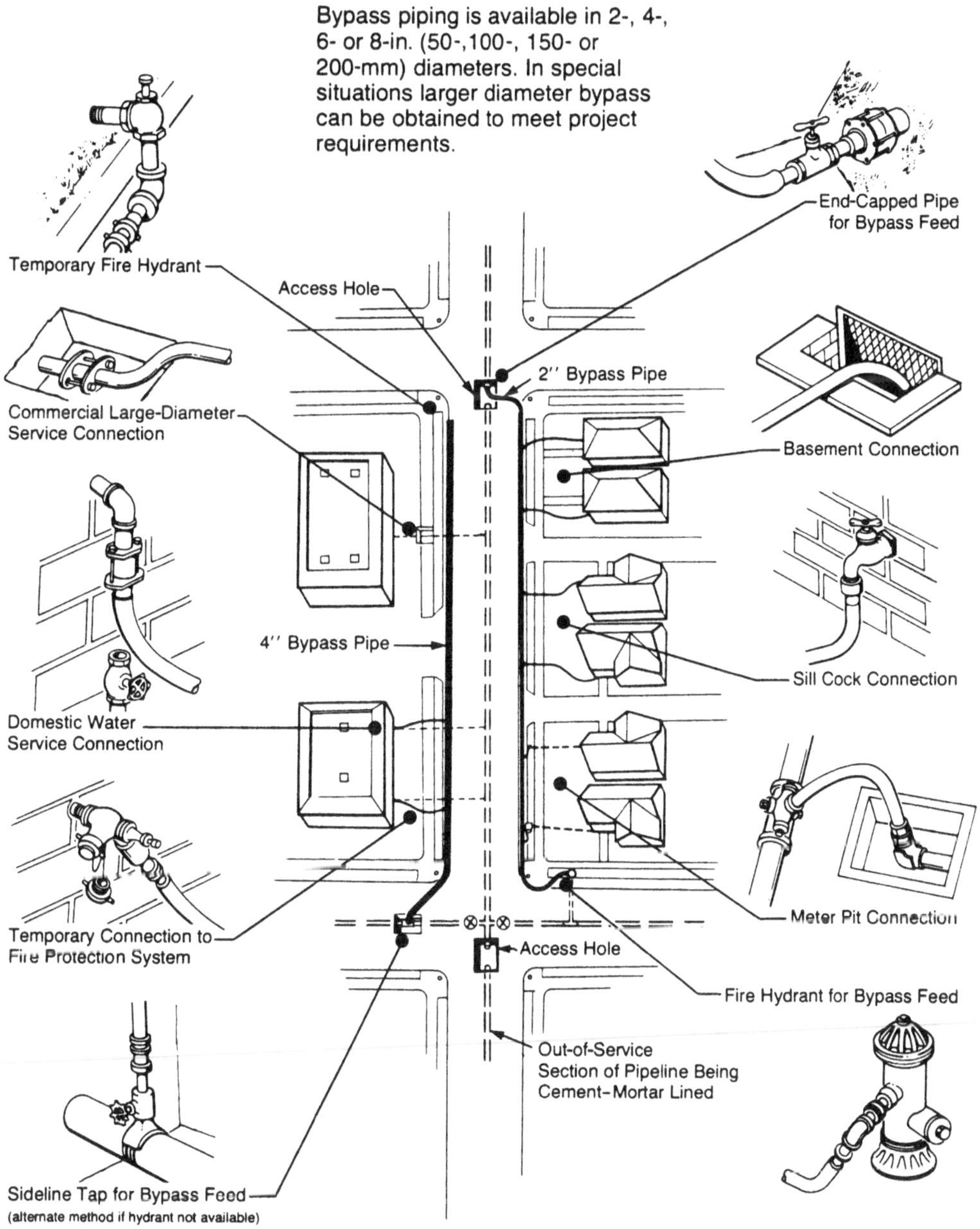


Figure 3-1 Bypass installation for residential and commercial water service

## COMMUNITY RELATIONS

Prior to the installation of a temporary water system, the project team and contractor should agree on a specific plan for notifying government agencies, emergency service providers, and their customers of an upcoming project. The plan should include the streets to be bypassed, the location of the feed hydrants, and the size and location of the bypass piping. Notification to the general public can be via newspaper articles, social media, utility bill mailings, or the utility's website; however, individual door hangers are recommended to notify the affected customers. These door hangers should be provided 24 to 48 hr in advance and should include contact information for general questions and emergency contact numbers as well. Refer to chapter 13 for more information on community outreach.

In addition to contacts with individual customers, overall community relations are equally important. Police and traffic control personnel and transit district representatives must be included in planning meetings to discuss detours and work schedules. The fire department also must be informed of the project and any effects on emergency access to all fire hydrants in the project area. Hydrants removed from service should be covered or otherwise marked to indicate their unavailability.

Other utilities, such as sewage, gas, electric service, and telecommunications (telephone, cable, and fiber-optic cable) should also be notified to allow coordination between the water line project and any other scheduled utility work. The local chamber of commerce or other local business groups may be contacted for assistance. Notices posted in store windows may help to avoid parking complaints.

## SUMMARY

All notifications should include an emergency telephone number that is active 24 hours a day, 7 days a week. Good customer relations can be severely damaged by a customer's inability to report an emergency or discuss questions about water service. If a project requires entering private property to disconnect meters and hook up temporary lines, notices should clearly discuss the need for the work and how it will be conducted. As with any aspect of customer service, follow-up is extremely important and personnel must be available to respond promptly to all calls. Finally, an established policy and procedure for handling damage claims should be established.

## REFERENCES

American National Standards Institute (ANSI). 2012. Drinking Water System Components—Health Effects. NSF/ANSI 61-2012. 198. Ann Arbor, Mich.: NSF International.

American Water Works Association (AWWA). 2005. Standard for Disinfecting Water Mains. ANSI/AWWA C651. Denver, Colo.: AWWA.

Chapter 4

# Pipeline Cleaning Methods

Water mains can be cleaned in place in a water distribution system using one or more of the following methods:

- Flushing
- Air scouring
- Mechanical cleaning techniques
- Cable-attached devices
- Fluid-propelled devices
- Power boring
- Ball cleaning
- Other methods

Dispose of cleaning water and debris in accordance with applicable regulations. Refer to ANSI/AWWA Standard C651, Disinfecting Water Mains, for guidance in disinfecting pipelines before returning them to service.

## FLUSHING

Water main flushing has a definite place in distribution system operation. It is particularly effective for light cleaning jobs. Mains should be flushed to clean newly installed and repaired mains prior to and after disinfection. Routine flushing is needed to remove impurities that cause complaints or are considered hazardous to public health and has become a vital part of maintaining the quality of water throughout the distribution system. Refer to the AWWA handbook titled *Disinfection of Pipelines and Storage Facilities Field Guide* for specific flushing program guidelines.

Flushing should never be considered the only solution to water quality problems identified in the distribution system. Proper design and operation of distribution facilities and an effective backflow-prevention program should also be part of any effort to maintain water quality. While quality control is the primary purpose of flushing, careful observation of system hydraulics during flushing may indicate problems in mains, such as inadequate capacity, undiscovered restrictions, or closed or partially closed valves.

All flushing and the disposal of the water should be done in accordance with federal, state or provincial, and local requirements.

## AIR SCOURING

The use of pressurized air to clean water distribution mains is not one of the more common processes. The sections of main to be cleaned have to be isolated to reduce the problems associated with the procedure. If the line to be air scoured has service connections on it, they will have to be isolated as well or placed on a bypass water delivery system so that the customers remain in service during cleaning. Air scouring is more effective for removing film and light-weight debris from the interior of the pipe by forcing high pressure air at a high velocity.

## MECHANICAL CLEANING TECHNIQUES

### Cable-Attached Devices

Systems that use cable-attached devices for cleaning distribution mains include drag cleaning, hydraulic jet cleaning, and electric scraper cleaning. In each case, the length of hose or cable determines the length of the pipe section that can be cleaned.

#### *Drag Cleaning*

In drag cleaning, a winch pulls a mechanical cleaner composed of a series of steel scraper blades and rubber squeegees through the pipe (Figure 4-1). The mechanical cleaner is usually flexible, allowing it to negotiate bends of up to 45°. Both ends of the cleaner are fastened to steel cables, which are attached to winches at either end of the pipe section to be cleaned. The cleaner is winched first in one direction and then in the other, and this process continues until the pipe is satisfactorily cleaned.

This cleaning method offers important advantages:

- Cleaning can be accomplished when water pressure or volume is insufficient to propel a hydraulically driven device, or when excessive pressure would be required for hydraulic cleaning, especially with small-diameter mains.
- Very hard deposits and encrustation can be removed.
- Dry solids allow easy disposal.

#### *Chain Scraper*

A combined hydraulic and mechanical tool called a *rotating chain scraper* or *reamer* may also be used to remove more difficult deposits from the pipe surface. This process incorporates the flailing of the chain on the interior of the pipe to dislodge encrusted material. The accompanying jet stream of water then flushes the particles outside the pipe to a point where they are disposed of. This provides a more aggressive method of cleaning the pipe while reducing the possibility of damaging penetrating service connections.

Figure 4-1 Drag cleaning, in which a winch pulls a mechanical cleaner through the pipe

### Hydraulic-Jet Cleaning

To clean pipe with a hydraulic jet, a special nozzle attached to a hose emits a jet of water at high velocity and pressure that removes debris and deposits from the interior of the pipe. Nozzle pressures of 1,000 to 10,000 psi (6,900 to 69,000 kPa) and above can be generated using this system. The jetted stream of water must bear hard enough against the scale or deposit to breach and dislodge particles. Once the scale or depositional material is penetrated, the fluid forms a wedge between the deposit and the surface, and strips off the deposit, exposing the clean metal surface. In many jetting operations, dislodged particles are entrained in the jetting stream and effectively dislodge more particles. The principal advantage of this method is removal of very tough deposits.

Hydraulic-jet cleaning uses a large volume of water and is more effective in larger diameter mains and conduits.

### Electric Scrapers

Electric scrapers are used to clean large diameter lines. A variation of this cleaning method is the use of an electrically powered, operator-driven scraper. This power-driven scraper incorporates revolving brushes or rotating arms to clean the line. The principal advantage of this method is the ability of the operator to evaluate the effectiveness of the cleaning process as it proceeds through the line.

## FLUID-PROPELLED CLEANING DEVICES

This section of the chapter describes fluid-propelled cleaning devices such as foam pigs and mechanical metal scrapers.

### Foam Pigs and Swabs

Foam pigs are flexible, bullet-shaped cleaning tools manufactured of high-quality (1.5 to 2.0 lb/ft$^3$ [24 to 32 kg/m$^3$] 5 to 8 lb/ft$^3$ [80 to 128 kg/m$^3$] density) polyurethane foam, some coated with polyurethane synthetic rubber coatings (70 lb/ft$^3$ [1,121 kg/m$^3$] density). (See Figures 4-2 and 4-3.) The sizes of the pigs range from 2-in. through 60-in. Pigs are propelled down water mains by the pressure and volume of water in the distribution system. Cleaning is accomplished by the frictional drag and flexible characteristics of the foam pig, which removes foreign objects, iron tuberculation, and other matter as it passes through the pipes, leaving the interior surfaces smooth and free from irregularities as age, attrition of use, and type of piping material will allow. When water pressure is applied for propulsion, a certain amount of water bypass (approximately 10 percent) helps to keep loose debris suspended in front of the foam pig (Figure 4-4).

Cleaning of deteriorating mains generally requires a series of swabs and foam pigs applied in progressively larger diameters and often with wire brushes attached (referred to as *progressive pigging*) until the pipe is restored to its original diameter.

Figure 4-2 Foam pig: A bullet-shaped device made of polyurethane foam

Figure 4-3 A foam pig with hardened coatings

Figure 4-4 Loose debris flushed ahead of the pig

## Operating Procedures

The following steps briefly describe the most common operating procedures for cleaning water distribution mains with foam pigs. The described procedure is based on a best-case scenario; actual procedures may vary for certain applications.

1. Review drawings of lines to be cleaned to identify (a) possible entry and exit points, and (b) all valves to be used to isolate the section of system to be cleaned.
2. Exercise all valves in the section to be cleaned before the pigging operation to ensure isolation of the section to be cleaned.
3. Flow tests should be performed before and after the pigging operation to evaluate the efficacy of the cleaning and determine the resulting condition of the water main.
4. Provide adequate means for disposing any debris exiting the pipe. The disposal of debris and the water used should be performed in compliance with local, regional, and governmental requirements. However, the best arrangement for handling large amounts of heavy debris would be to direct the flow to a temporary settling pond, the bed of a dump truck, etc., to facilitate removal of solids once the water has drained away.
5. Notify all affected customers (commercial and residential) as well as fire department officials of the scheduled interruption of service.
6. Confirm that all valves in the section to be cleaned are fully opened and working properly and that the section is properly isolated.
7. Follow the direction of flow, which must be controlled by the valve locations.
8. Introducing the foam pigs into the water main may be accomplished either by hand or through the use of mechanical equipment commonly called a *pig launcher*.

Several types of fire hydrants, once they are disassembled, can be used as entry and exit points for pigging of some water mains 6 in. (150 mm) and smaller (Figure 4-5).

9. To produce the best cleaning results, foam pigs should travel through a water main at a rate of 3 to 9 ft/sec (0.91 to 2.74 m/sec).
10. Sequence of foam pig runs:
    a. Introduce a line-sized, soft-foam swab (1.5 to 2.0 lb/ft$^3$ [24 to 32 kg/m$^3$] density, uncoated), called a *prover pig*, to determine the actual effective inside diameter of the pipe to be cleaned. The swab wears down while passing deposits in and along the pipe wall.
    b. Introduce a line-sized bare foam pig (5- to 8-lb/ft$^3$ [80- to 128-kg/m$^3$] density, uncoated) to remove soft deposits and help gauge the true opening in the line.
    c. Introduce a line-sized, coated foam pig (5- to 8-lb/ft$^3$ [80- to 128-kg/m$^3$] density, coated with urethane elastomer). This type of foam pig should be run repeatedly until one emerges in reusable condition. A water line with excessive buildup should be pigged using the progressive pigging method, beginning with undersized, coated foam pigs and gradually increasing the size in successive passes until a line-sized foam pig emerges in reusable condition.
    d. Line-sized, wire brush foam pigs may be introduced following these applications for final removal of extremely hard deposits (such as tuberculation).
    e. Introduce a line-sized swab (1.5- to 2.0-lb/ft$^3$ [24- to 32-kg/m$^3$] density foam, uncoated) to sweep out any loose debris and determine the effectiveness of the cleaning process.
11. Flush in accordance with requirements and disinfect as required.
12. Place the water main back into service.

Figure 4-5 Pigs launched through a disassembled fire hydrant for a 6-in. (150-mm) or smaller line

Foam pigs are flexible enough to negotiate short-radius bends and pass through full-port valves and some types of plug valves. Their capability to be compressed up to 35 percent of their cross-sectional area allows them to pass through multidimensional lines and full-port valves. A piggable valve is one through which a pig will pass every time. Foam pigs do not pass through butterfly valves. Foam pigs can clean long sections of piping from one entry point, thus reducing the number of excavation sites required. Pigging is a fast, simple, and cost-effective method of returning most deteriorated piping systems to as-designed capacity and efficiency.

## METAL SCRAPERS

A metal cleaning scraper consists of a steel frame shaped like a piston. Specially tempered steel blades are attached around the scraper at various angles to create a scraping and brushing action (Figures 4-6 through 4-8). The cleaner is propelled through the water main using water pressure.

Cleaning is often accomplished with a single pass in a continuous operation; sometimes, however, interior pipe conditions may require additional passes. The length of pipeline that can be cleaned hydraulically in one operation is limited only by the availability of volume and water pressure and a proper means of disposing of water and deposits. An opening must be provided at each end of the section to be cleaned for entry and exit of the cleaning tool. The volume of water required to hydraulically clean a pipeline will depend, to a great extent, on how dirty the water is as well as the amount of tuberculation build up inside the pipe.

Figure 4-6 Scraper unit with specially tempered steel blades

Figure 4-7 Several scraper units assembled together in the field

Figure 4-8 A series of disks to act as a hydraulic piston, pulling scrapers through the line

Sufficient water must be added behind the cleaner to fill the pipe as the cleaner moves ahead. The water that passes the cleaner scours the wall of the pipe and washes ahead the material that is scraped off the pipe. While the velocity of water ahead of the cleaner is independent of cleaner speed, it must be sufficient to remove the deposits. Experience indicates that a flow velocity ahead of the cleaner of between 2 and 10 ft/sec (0.6 and 3.0 m/sec) is required to remove the deposits. In a small pipe, because a relatively large amount of material must be moved for the volume of the pipe, a relatively high velocity is required.

The cleaning water and deposits must be discharged from the pipeline in a way that avoids creating a nuisance or environmental problem and in accordance with the requirements of regulatory agencies. A sandbag dam can be constructed to create a pond for particle settlement (Figure 4-9). This allows the clean water to be decanted, dechlorinated, and disposed of in accordance with local regulations while any solid material is collected and properly disposed of. Another method may be to pipe the water to a dump truck, catching the solids in the truck.

Figure 4-9 A sandbag dam to create a pond for particle settling

## Operating Procedures

The following steps describe the operating procedures commonly used when cleaning water pipes with mechanical scrapers.

1. Review drawings of lines to be cleaned. Identify logical entry and exit points for the cleaning tool. Note all valves, service connections, and other components that must be opened or closed to isolate the line to be cleaned. As described in chapter 1, conduct flow testing of the pipeline to determine the Hazen–Williams *C* factor (if not previously completed) to determine the condition of the line.
2. Excavate and install a spool piece at each entry and exit point (Figure 4-10). Obtain a sample of deposits in the line and determine their average thickness.
3. Determine source of cleaning water (e.g., reservoir, feeder mains, parallel lines, etc.). Cleaning for lines 12 in. (300 mm) in diameter and smaller may require auxiliary pumps to provide adequate flow and pressure; flow and pressure within the system generally are adequate to clean mains over 12 in. (300 mm) in diameter.
4. Provide suitable means for disposing water and removed solids that meets with the regulatory requirements.
5. Notify all affected commercial and residential customers, as well as fire department officials, of the scheduled interruption of service.
6. Completely isolate the section to be cleaned. Close valves slowly to prevent water hammer.

7. Drain the line to be cleaned.
8. Open entry and exit points and remove spool piece sections. At the entry point, install a spool piece with the cleaning device inside. At the exit point, prepare a method to remove discharge water. Connect a water manifold with a throttle valve to control the rate of travel of the cleaning device.
9. Initiate the required flow into the main, and control speed with a bypass, blowoff valve.
10. Complete the cleaning operation.
11. Reclaim the cleaning device at the exit spool.
12. Repeat additional cleaning passes, as required.
13. Remove the insertion and retrieval sections, and replace permanent spool sections. Backfill and complete other necessary maintenance.
14. Flush in accordance with requirements and disinfect the line, as required. Conduct flow tests to determine the cleaned pipeline Hazen–Williams *C* factor following cleaning.
15. Return the line to service.

Figure 4-10 A spool piece installed at the entry and exit points for mechanical scrapers

### Advantages

Principal advantages of this method include

1. Ability to clean long stretches of heavily deposited pipe at 2 to 10 ft/sec (0.6 to 3.0 m/sec) in a single, continuous operation
2. Flexibility to negotiate standard bends (with maximum radius equal to 1½ pipe diameters) and elbows, inclined and vertical pipe sections, and to pass line-size gate valves, ball valves, tees, and corporation taps
3. Ability to clean water mains with minimal excavation points
4. Capability to restore water mains to a Hazen–Williams *C* factor comparable to a newly installed, unlined pipe. Refer to chapter 1 for the discussion regarding the use of Hazen–Williams equation.

## CLEANING BY POWER BORING

Power boring is a cleaning method using any hydraulically powered device capable of removing tuberculation and encrustation from cast-iron, ductile-iron, and steel pipe from 3 in. (76 mm) in diameter and above. This process is normally carried out in lengths of 400 ft (121 m) or greater.

A rack-feed boring machine (Figure 4-11) is a compact, diesel-powered unit that uses hydraulic pressure to deliver up to 31 hp (23.1 kW) to a boring head. The boring head is designed to accommodate spring steel boring rods 15-ft (4.6-m) long fitted with spring-loaded quick-connects for connecting rods into suitable lengths for cleaning various lengths of pipe (Figure 4-12). The end of a boring rod assembly is fitted with a spring steel cutter blade or other cleaning tool, which rotates at 750 rpm through the pipe (Figure 4-13) This cleaning process is conducted against a controlled, upstream water flow to flush loosened debris from the pipe.

The rack-feed boring machine may be equipped with an adjustable boom to accommodate various pipe depths and to control the angle at which boring rods are inserted into the pipe. The ratio of boring rate to spring cutter blade revolutions is predetermined and fixed to eliminate operator error. This setting ensures consistent results throughout the cleaning operation.

The rack-feed boring method leaves a pipe's interior surface free from tuberculation and encrustation and can be effective for bends up to 22.5°. Bends of greater radius may require removal and replacement.

Figure 4-11 Rack-feed boring machine

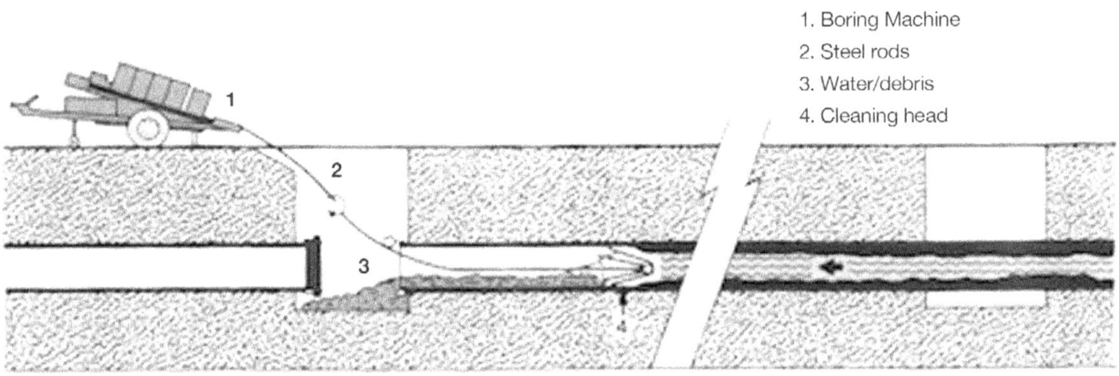


Figure 4-12 Cleaning pipe by power boring

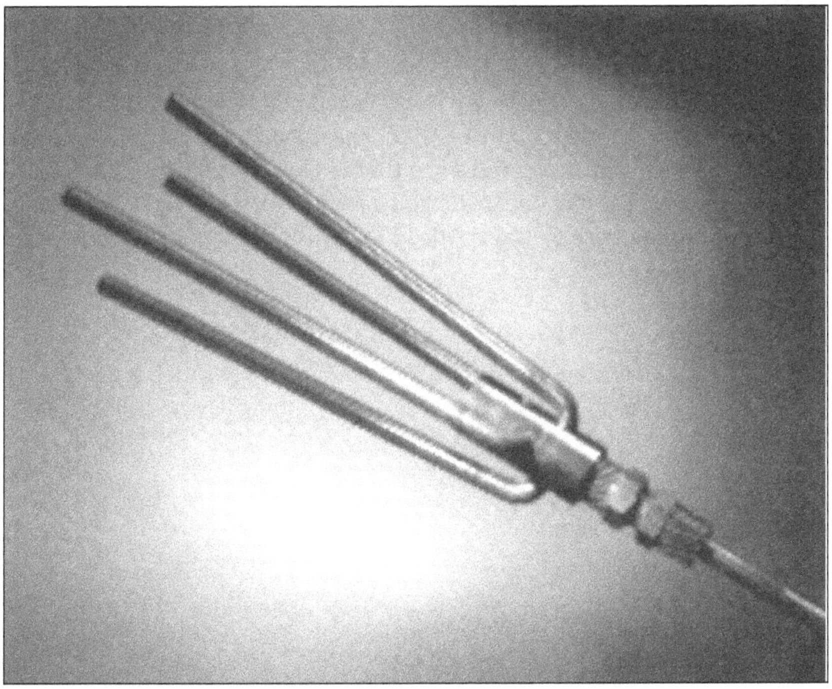

Figure 4-13 Cleaning head

## Operating Procedures

The following steps briefly describe the basic operating procedures:

1. Review the drawings of lines to be cleaned and verify by site inspection all possible entry and exit points and the locations of valves, fittings, etc. Use pipe-locating equipment and clearly mark buried utility services.
2. Coordinate any remedial action with the utility.
3. Conduct flow tests (if required) before and after cleaning to determine the condition of the main.
4. Notify all customers of work to be conducted.
5. Lay temporary service, chlorinated in accordance with established standards.
6. Access all points of entry in accordance with contractual documents.
7. Following the boring machine manufacturer's operating instructions, bore clean the pipe. Capture all debris for proper disposal.
8. Following bore cleaning, remove residual water from the pipe using foam swabs and/or rubber squeegees and dispose of in accordance with regulatory requirements.
9. If available, color closed-circuit television equipment may be used to inspect the pipe and produce a video, noting street names and addresses inspected.

## BALL CLEANING

This method of removing deposits from the interior of water mains utilizes balls propelled through the line. These rubber balls are placed in the distribution main and forced through the main under the pressure of water. These balls are directed from the entry point to a location where a strainer retrieves them. The random movement of the resilient balls through the pipe loosens the encrustations and forces them to an exit point downstream. At the end of the run a strainer is used to collect the balls and the debris to be discarded.

## REFERENCES

American Water Works Association (AWWA). 2005. Standard for Disinfecting Water Mains. ANSI/AWWA C651. Denver, Colo.: AWWA.

AWWA. 2006. *Disinfection of Pipelines and Storage Facilities Field Guide.* Denver, Colo.: AWWA.

Chapter 5

# Cement–Mortar Lining

## CEMENT–MORTAR LINING

When alkaline-containing water comes in contact with iron, a chemical inhibitor against oxidation forms. Because the cement and sand in cement–mortar lining is porous, water can penetrate through the lining to the pipe wall, becoming alkaline in the process. Consequently, pipe that is lined with cement mortar is protected from oxidation because of the composition of the Portland cement.

Cement–mortar linings were first installed in existing pipelines using the centrifugal process in the mid-1930s. However, this method was limited to pipelines large enough for a person to enter. In the 1960s, remote lining processes were introduced. Currently, cement mortar is applied to new ductile-iron pipes and most new steel pipes before installation, making this method a standard in the water industry.

Cement mortar is applied to the pipe wall by the rotating head of an electric or air-powered machine. The machine is placed in the pipe at 300- to 1,500-ft (92- to 458-m) intervals, depending on pipe diameter, valve locations, bends, profile, and alignment (Figures 5-1, 5-2, and 5-3). Mortar is pumped to the lining machine through high-pressure hoses or is mechanically delivered. The lining machine is equipped with rotating trowels or a conical drag trowel positioned just behind the dispensing head. As the machine moves through the pipe, it leaves a smooth, troweled (nonstructural) finish.

A reinforced cement–mortar lining may also provide structural improvement. Initially a 0.5-in. (13-mm) thick cement–mortar lining is placed on the pipe wall in the conventional manner without troweling. Using overlapping joints, a wire mesh is placed against the lining, which is then covered by another 0.5-in. (13-mm) troweled cement–mortar lining. The remaining steps in the process are the same as those for an unreinforced lining. This process is generally used to rehabilitate steel lines in poor condition due to extensive corrosion and in lines large enough for a person to enter. The reinforcing wire holds the cement mortar together, even if large holes develop in the pipeline. While this method will not necessarily prevent a blowout, the useful life of a pipeline will be extended.

Figure 5-1 A cement–mortar lining machine for use in small-diameter pipe

Figure 5-2 Introduction of a small lining machine

Figure 5-3 A cement–mortar lining machine for use in large-diameter pipe

## Operating Procedures

The following steps describe the common operating procedures used when lining a main with nonstructural cement mortar.

1. Topography, bends, in-line valves, reduced-diameter in-line valves, and traffic conditions all influence the locations of excavations for access points to allow introduction of the lining equipment. These conditions also affect the method used to clean the pipe. All 22.5°, 45°, and 90° bends must be removed for 12-in. (300-mm) diameter pipe and smaller; 45° and 90° bends must be removed in 16-in. (400-mm) diameter pipe. Pipe 20 in. (500-mm) diameter and larger may require excavations near the bends or removal. Full-diameter line valves must be removed and replaced, or the bonnets removed to clean the cement mortar from the interior of the valve if the pipe is too small for a person to enter. In large pipe, personnel can clean the valve interior from inside the pipe. Depending on their size, reduced-diameter or obstructed line valves may require removal or excavations on each side of the valves to allow lining.
2. The pipeline is generally dewatered by gravity and/or blowoff valves, which remove most of the water. The contractor can then remove any remaining water lying in low spots with rubber squeegees pulled by a winch through the pipe. At this time, inspection usually determines whether any branch valves are leaking. If a valve fails to stop water from entering the pipe to be lined, it is re-exercised. If the leakage continues, a seal is installed or the valve is replaced. The pipe must be free of water when lining with cement mortar.
3. Cast-iron pipe is cut with a squeeze-type cutter, guillotine saw, handheld mechanical saw, or traveling saw, and a 4- to 5-ft (1.0- to 1.5-m) section of pipe is removed.

Steel pipe is cut by torch, and a half cap is removed. Mechanical couplings are used for reconnecting cast-iron pipe, and butt straps are used to reconnect steel pipe.

4. A mixing van or concrete batch plant is located near the access hole where a cement–mortar mixture, consisting of one part silica-type sand and one part type II Portland cement, is prepared. The mixed mortar is then delivered to the lining machine by one of several methods, depending on the pipe diameter.
5. In 4- to 24-in. (100- to 600-mm) diameter pipe (sometimes larger), the mortar is pumped through high-pressure hoses to the lining machine. Specially designed winches pull the lining machine through the pipe at a constant speed, ensuring a uniform lining thickness.
6. Mechanical feeding equipment shuttling between the access excavation (where the mortar is mixed) and the lining machine delivers mortar to the lining-machine hopper. The lining-machine operator then regulates the mortar application. Mechanically driven rotating trowels are used with this manually operated equipment.
7. Consumers' service lines and laterals less than 2 in. (50 mm) in diameter must be cleared after the lining application. This is done about 1 hr after the lining is completed, using compressed air to blow open the service line at the connection to the main. Laterals over 2 in. (50 mm) in diameter are not plugged by centrifugal lining and do not require excavation or blow back.
8. Water can be introduced into the line without pressure to allow curing 24 hr after completing the lining. The main can then be chlorinated, tested, and returned to service. (Discharge water should be disposed of in accordance with governmental or local ordinances. Permits may be required for disposal of the water. Chlorine may need to be neutralized and the pH adjusted before discharge. Mains with low flow may experience high pH problems for a short time after being returned to service. Most distribution mains are returned to service between four and seven days after lining (Figure 5-4), depending on valve locations and disinfection requirements.

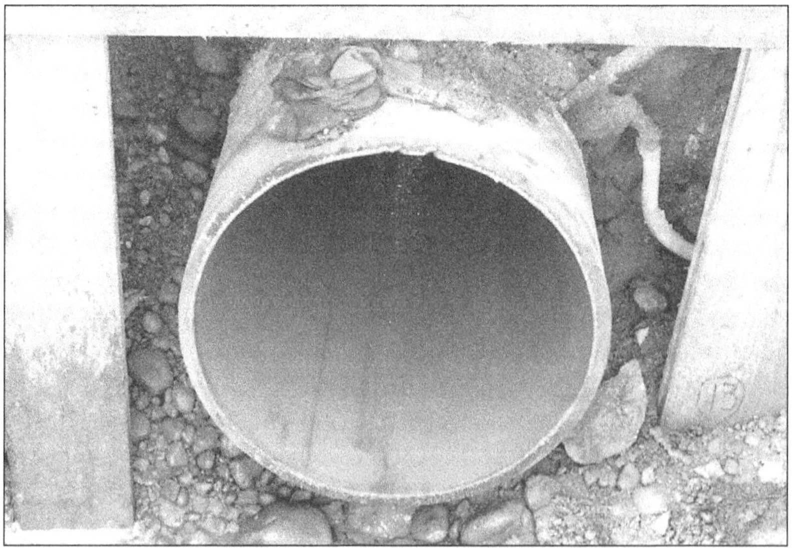

**Figure 5-4 A pipe ready to be returned to service four to seven days after cement–mortar lining**

## REFERENCE

American Water Works Association (AWWA). 2011. Standard for Cement–Mortar Lining of Water Pipelines in Place—4 In. (100 mm) and Larger. ANSI/AWWA C602. Denver, Colo.: AWWA.

This page intentionally blank.

Chapter 6

# Spray-On Polymer Lining

The process for in-situ renewal of iron and steel pipelines with sprayed polymeric materials was developed in the United Kingdom using two-part epoxy material in the late 1970s and has been performed in North America since the early 1990s. The process has been used effectively to rehabilitate old, unlined water mains that are in a structurally sound condition. Epoxy lining was first demonstrated in the United States in 1992. The epoxy materials approved for use were first certified by ANSI/NSF Standard 61 in 1995, and an American Water Works Association (AWWA) standard for epoxy lining was issued in 2008.

A new generation of materials, polyurethane and polyurea-based fast-setting polymer linings, were developed in the United Kingdom. Since the first commercial application of spray-applied polymers in the United Kingdom in 1989, they have become increasingly popular for many applications formerly dominated by epoxy. A primary driver for this is the quick cure characteristics that allow cleaning, lining, and return to service of the water main in the same day when supplemented with superchlorination or alternatively, with boil-water advisories.

These polymeric compounds are now available in the United States and Canada, and are rapidly expanding into many other countries. These materials offer a rapid cure time—to as little as 30 min/7 sec (7 sec tack free) versus 12–16 hr for epoxy. The use of epoxy and cement–mortar lining for renewing water mains has generally been considered a method to install a simple barrier coating. Some recent polymer-based materials offer the ability to structurally enhance the host pipe and extend operational life. Some materials add structural value by bonding to the host pipe while other products offer independent structural support.

## DEFINITION OF POLYUREA MATERIALS

The epoxy resin most widely used is made by reacting epichlorohydrin with a compound such as diphenylol propane. Polyurea is formed from the reaction of isocyanate and an amine. Polyurethane is formed from the reaction of isocyanate and polyether polyols. Linings can also be formulated as hybrids by combining isocyanate with a mixture of polyols and amines, resulting in a material that bears the performance characteristics of both polyurethane and polyurea.

Polymeric coatings can renew or enhance water pipes and provide the following features:

- Eliminate pipeline internal corrosion and water quality issues associated with internal corrosion: Spray-applied polymeric linings provide a highly effective and corrosion-resistant barrier. This barrier eliminates tuberculation, which contributes to deteriorating water quality.
- Typically increase flow capacity in ferrous pipe: Spray-applied polymeric linings present a smooth surface to help maximize the flow capacity of the pipe. Flow improvement is due to the low friction coefficient and the relatively thin-wall thickness of the lining. The lining itself is generally thinner than cement lining and slip-lined alternatives.
- Offer faster return to service: Polymer-based spray-applied linings permit a rapid cure capability that can dramatically reduce shutdown periods and potentially provide same day return to service of water lines, eliminating the need for bypass piping subject to prevailing requirements for disinfection and bacteriological testing. A two-stage operation is possible in one or two days. For the two-day process the pipe is cleaned but not lined, disinfected, and reconnected at one end to restore service after the first day. On the second day, the pipe is disconnected, dewatered, (dried if needed), inspected, lined, disinfected, flushed, and reinstated. For a one-day process, the pipe is cleaned, inspected, lined, inspected again, and returned to service. Accelerated return to service is highly dependent on the rapid cure time and the applicable criteria for disinfection and bacteriological testing.
- Do not contain volatile organic compound (VOCs): Two-part spray-applied polymeric linings should not contain any VOCs from added solvents, which can be a source of air pollution and present problems with local air quality ordinances. This should be validated with the material supplier.
- Avoid plugging of service connections: During installation, spray-applied polymeric linings generally do not close or plug the pipe service connections. However, robotic reinstatement of service connections may be needed if they are undersized relative to the pipe diameter or if thicker structural linings are installed on small diameter pipelines.
- Accommodate pipe bends: Spray-applied polymeric linings can be applied through bends up to 45° avoiding the cost of opening and removing the bends through multiple excavations. The application can also work past large service connections and hydrant tees without causing damage.

The polymeric materials that meet AWWA Class II or III semistructural criteria have these added features:

- Potential reduction of leakage from corrosion holes, cracks, and failed pipeline joints: Spray-applied polymeric linings provide a continuous envelope inside the existing host pipeline and can be formulated to span minor corrosion holes and minor joint gaps.
- May improve structural integrity and design life of old pipe: Some polymer linings may add minor strength to the host pipe. These are identified as a Class II semistructural lining. Polymer-based spray-applied linings can be formulated with sufficient ring stiffness to provide interactive pressure support for the host pipe without relying on adhesion. This is considered AWWA Class III semistructural per this manual.

To meet AWWA Class IV structural criteria, polymeric material must have the ability to essentially replace the host pipe in the event of structural failure, and continue to perform on a long-term basis. Should the host pipe fracture, a Class IV spray-applied lining must separate from the host pipe much as Class III materials but have sufficient structural strength to function as an independent pipe under load and full working pressures. At the time of publication, there are no conclusive tests that demonstrate this ability for a commercially available spray-applied lining.

Like other trenchless technology methods, spray-applied polymeric linings tend to minimize social costs and disturbance to adjacent utilities and structures, as well as minimizing surface and subsurface excavations. With conventional open-cut construction methods, direct costs are greatly increased by the need to restore ground surfaces such as sidewalks, pavement, and landscaping. Additionally, factors related to open-cut methods include adverse impacts on the community, businesses, and commuters due to air pollution, noise and dust, safety hazards, and traffic disruption.

Benefits of trenchless renewal using spray-applied polymeric linings are summarized in Table 6-1.

**Table 6-1 Spray-applied polymer lining features and benefits**

| Features | Benefits |
|---|---|
| Continuous barrier lining | Stops and prevents internal corrosion of old pipe<br>May prevent leakage from pipe joints and holes |
| Thin-walled lining | Maximizes inside diameter of the relined pipe |
| Close-fit lining | Maximizes flow capacity. Grouting is not required. |
| Smooth surface | Maximizes flow capacity |
| Ambient temperature installation | No process heating requirements |
| Trenchless installation | Minimum disturbance to adjacent services, surface and subsurface, and structures. Less environmental and social disruption compared with open-cut pipe construction. |
| Simple and flexible installation equipment | Applicable to pipe diameters from 4 in. (100 mm) and greater |
| Return to service | Same-day return to service capability when using quick-cure materials depending on local disinfection requirements or boil-water advisories |
| Abrasion resistance | Provides long-term abrasion resistance |

Spray-applied polymeric linings are based on mixing of two-component starting materials. It can be specified for use in water pipeline renewal applications when shown to have ANSI/NSF Standard 61 Drinking Water System Components certification. The starting materials generally have low viscosity characteristics to enable pumping through an umbilical hose to the spray head, and high build slump resistant linings with excellent adhesion characteristics can be obtained. Finished linings are hard, glossy, and free of surface tackiness or greasiness.

Nonstructural spray-applied polymeric linings require a uniform internal surface to avoid discontinuities at joint areas, thus ensuring optimum thickness of linings throughout the length of the application. The speed of application results in rapid renewal of water pipes with quick return to service.

Renewal operations usually require service disruptions and may include a significant amount of equipment or personnel in residential and business districts. However, with proper and realistic planning and scheduling, spray-applied polymeric linings will allow cleaning, inspection, application, and reinstating of the water service in minimal time.

In planning a project, it is essential to locate all valves and hydrants along the length of the pipe to be lined. It is recommended that all dimensions in the form of drawings, videos, etc., be checked by the contractor/applicator to ensure accuracy. Once this is achieved, the number of sections, pit locations, and the lengths of the linings to be installed can be determined.

Lengths of about 660 ft (200 m) of pipe can be lined between two access pits. Spray direction can be changed to allow uniform coating when severe misalignments are observed, i.e., displaced joints, diameter changes, etc.

Prior to lining, it is strongly recommended that each pipe segment be assessed by an up-to-date internal inspection using a closed circuit television (CCTV). This inspection will indicate the extent of cleaning required, highlight any defect(s) that must be taken into account when considering the appropriate lining thickness, locate active and abandoned service connections, and identify a level of moisture that cannot be tolerated and any active source of water that might still be entering the pipe.

Pipe cleaning and surface preparation: In order to achieve proper bonding and adhesion to the surface of the host pipe, the interior of the pipe must be properly cleaned before the lining is applied. There are numerous cleaning methods, each of which have their respective advantages and applicability depending on required adhesion properties for the polymer product to the host pipe (see chapter 4). Air-surging and swabbing can be used on clean pipes to remove dust and other minute debris. To be certain that pipe has been cleaned adequately and that free-standing water has been removed and leaks have been stopped, a CCTV inspection should be performed prior to spray lining.

The equipment used for spray-applied polymeric lining application should be suitable for storage, heating, dispensing, and mixing and should be in accordance to the lining manufacturer's instructions. Lining equipment should be fitted with flow meters with subsidiary pressure monitors and must be capable of dispensing and monitoring the two components (base and activator) within manufacturer's tolerance (usually ±5%) of the specified mix ratio. The equipment should be fitted with an audible alarm, which is activated when the mix ratio departs from ±5% of that specified.

The application of the proper mix and amount of product employs microprocessor technology to control delivery of the two materials, temperature control, and travel speed of the spray equipment. Lining equipment provides a continuous record (display and hard copy) of the following as a minimum:

- Volume and flow rate of material delivered to the application head
- Mix ratio by volume
- Pressures in both base and activator hose lines
- Lining thickness
- Elapsed time from start of lining
- Date and real time
- Lining equipment should be capable of individually heating both base and activator components so that the material temperature at the application head is per the manufacturer's recommendation. In order to facilitate this, lining rigs shall be equipped with heated umbilical delivery hoses.
- On completion of the lining operation, the mixer hose and spray head should be cleaned by flushing with appropriate cleaners. All solvent washings should be removed from the site by the contractor for subsequent disposal as hazardous waste.
- Prior to commencement of the next lining operation, the mixer hose and spray head should be thoroughly purged with product during the spin-up process.
- Connections of new pipes or cut outs will require the old pipe and lining to be cut. In order to prevent de-bonding of the lining from the old pipe, cutting should be undertaken using a pneumatic reciprocating saw or similar. Use of such equipment avoids excessive heat buildup at the cut location, which could otherwise cause local distortion of the lining. End sealing of the lining to the host pipe may also be required or considered as per the lining manufacturer's recommendations.

Figures 6-1 through 6-3 show a typical set-up for sprayed-applied polymeric lining.

Courtesy of 3M, Inc.

**Figure 6-1 Lining spray in progress**

Courtesy of 3M, Inc.

Figure 6-2 Lining equipment vehicle

Courtesy of 3M, Inc.

Figure 6-3 Spray head retrieval pit

## Operating Procedure Overview

1. Locate excavation points, and evaluate in-line bends, valves, and other conditions to determine access locations.
2. Thoroughly and properly clean pipes, remove standing and leaking water, or dry as per manufacturer requirements.
3. Verify condition of the pipe to be lined using CCTV. Any structural holes or joint gaps that may be visible may require evaluation of a trained engineer and properly repaired/prepared prior to spray lining as required. Direct condition assessment of the host pipe prior to the project or after cleaning may identify otherwise nonvisual graphitization and remaining wall thickness and provide additional data that may be used to justify lining installation thickness.
4. Apply polymeric lining only when the pipe wall temperature is above 38°F (3°C) or minimum temperature specified by the material manufacture, whichever is higher.
5. Prior to lining, conduct checks on the equipment. Verify pump output, mix ratio, and material temperatures. Record this and other pertinent site information on a lining operation record sheet. Complete a separate record sheet for each separate lining run.
6. Prior to inserting delivery hoses into the water main, pump and circulate the two components until each component reaches the uniform operating temperature range specified by the material manufacturer.
7. Once the hose is inserted through the pipe section and the static mixer and application head are attached, check for proper operation. Visually check for correct mixing of the two material components by test spraying the mixed material through the spray head into a container and onto a test card and recording the observed mixed material color on the record sheet.
8. The application can begin when the operator is satisfied that the material flows are established and the lining material color is correct and uniform.
9. Carefully monitor the winch speed and the rate of withdrawal of the hoses to ensure (a) a smooth traverse, and (b) that the rate of withdrawal leaves a uniform thickness in a single application.
10. Immediately after completing application of the polymeric lining and inspecting the pipe, cap the ends of the main to prevent contamination and/or water from entering the pipe.
11. Allow the minimum cure period specified by the material manufacturer.
12. For some systems, a second lining layer is required. Apply per manufacturer directions repeating steps 5–12.
13. After the cure period, visually inspect the pipe at both ends and conduct a complete CCTV to verify the quality of the application. Typically the two-part product comes in two distinct colors that form a third color when applied to the pipe wall.
14. If the CCTV or visual examination reveals flaws, make corrections consistent with AWWA standard or material installation/repair guidelines.

The thickness of spray-applied polyurea linings can be varied between 0.040 in. (1 mm) to 0.340 in. (8.5 mm). The overall thickness of the polymeric product may be dependent on the desired structural capacity of the lining. Some systems utilize more than one coating to achieve the desired thickness for structural support.

## REFERENCES

American National Standards Institute (ANSI). 2012. Drinking Water System Components—Health Effects. NSF/ANSI 61-2012. 198. Ann Arbor, Mich.: NSF International.

American Water Works Association (AWWA). 2008. Standard for Spray-Applied In-Place Epoxy Lining of Water Pipelines, 3 in. (75 mm) and Larger. ANSI/AWWA C620. Denver, Colo.: AWWA.

Chapter 7

# Cured-In-Place Pipe Lining Techniques

Cured-in-place pipe (CIPP) lining techniques involve inserting a polymer fiber reinforced tube or hose impregnated or coated with a thermoset resin system into the host pipe. The resin is then cured, either under ambient conditions or by application of heat using steam or water, to produce a rigid "pipe within a pipe," capable of being custom designed as an interactive lining system or as a fully structural, independent lining system.

Techniques of this type have been used extensively for rehabilitation of municipal and industrial wastewater or pressurized pipelines and municipal gas lines. As with any lining technology used in potable water pipelines, applications are limited by the need to obtain appropriate certification or approval from relevant health authorities. Currently, several CIPP systems have been certified to NSF/ANSI Standard 61.

## CLASSIFICATION OF SYSTEMS

CIPP systems can be classified into three main groups:

1. Felt-based systems. The lining system is produced from a nonwoven polyester felt coated on the outer face with a layer of elastomer. For pressure pipeline situations, varying the thickness of the felt liner and/or including reinforcing fibers allows the lining, when impregnated with resin and cured, to meet a wide range of design requirements.
2. Woven hose systems. The lining tube consists of many circular woven, seamless, polyester fiber hoses coated on one outer face with a layer of elastomer and impregnated with resin. The composite construction of the hose and resin can be designed to carry varying internal pressures and external loads as required.
3. Membrane systems. The lining tube consists of a very thin elastomeric membrane designed only to offer internal corrosion protection and bridge very small pinholes

and joint gaps. The cured resin layer serves only as an adhesive to the host pipe. One variant joins such a membrane with a woven hose for a combination that can span larger holes and gaps.

Systems are also available based on combinations of these types.

## Felt-Based Systems

The original felt-based lining system, developed in the United Kingdom in 1971, has since been used to rehabilitate thousands of miles of wastewater and drinking water pipes throughout the world. The system has been utilized in pipe ranging from 4 in. to 108 in. (102 to 2,743 mm) in diameter, and it can accommodate noncircular shapes and negotiate 90° bends. A number of similar processes have evolved over the years.

Felt-based CIPP systems may vary depending on whether the lining system incorporates textiles or glass reinforcing fibers (Figure 7-1). The felt-based lining systems are manufactured to suit specific host pipe dimensions and impregnated with the appropriate resin, either in the contractor's factory or sometimes, for large-diameter pipes, on site. The resin used, either polyester, vinyl ester, or epoxy (typically used in potable water applications), is selected to meet mechanical performance requirements.

Felt-based systems can be classified according to the installation method. Some systems are installed by inversion, in which the impregnated liner is simultaneously fed through the pipe and turned inside out by water or air pressure (Figure 7-2). Some are installed by pulling a liner into a pipe and inflating it with water or air.

The impregnated liner is normally cooled in ice and transported to the job site in a refrigerated truck to prevent premature setting of the resin. The liner is then inverted into the pipe using pressurized water or air. Alternatively, it may be winched into the pipe and then inflated using air or water pressure. The liner is then cured by either heating the inversion water or using steam (Figure 7-3). After curing, the ends of the liner are cut flush with the end of the pipe. To ensure a sealed system at termination points in the lining system, as well as at reinstated service connections in distribution piping, some systems rely on the adherence of the liner resin system to the host pipe. In other cases, lining systems will rely on mechanical fittings to achieve a seal.

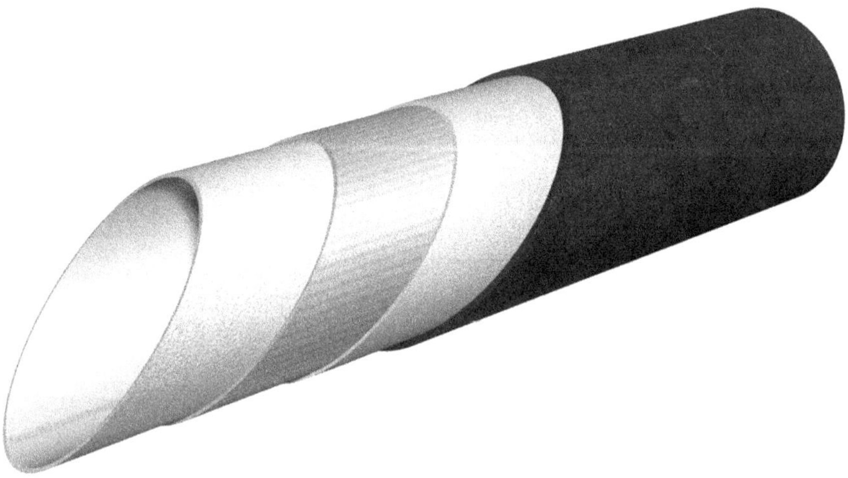

Courtesy of Insituform Technologies, LLC

**Figure 7-1 Glass reinforced CIPP felt composite lining system**

Courtesy of Insituform Technologies, LLC

Figure 7-2 Felt liner fed into the pipe utilizing the inversion process

Courtesy of Insituform Technologies, LLC

Figure 7-3 Curing of liner by heating of inversion water after liner is in place

Felt-based systems can offer from Class III semistructural capability to Class IV fully structural capability depending on its thickness and strength properties. Their use is indicated in pipes ranging from severe internal corrosion and pinhole leakage, leakage due to faulty joints, problems arising from localized external corrosion, and for severely deteriorated or structurally unsound pipelines. The installed CIPP liner is thin and achieves a tight fit with the host pipe. Its high Hazen–Williams value and joint-free construction may allow flow rates in the rehabilitated pipeline identical to those in the original pipe in new condition.

## Woven Hose Systems

These systems utilize circular woven polyester fiber hoses coated on one face with a layer of elastomer (Figure 7-4). Prior to installation, the hose is impregnated with a layer of epoxy resin and then pulled into the pipe using a winch or inserted in the pipe by an inversion technique using either air or water pressure.

In the first case, once the liner is pulled into place, the hose and resin are pressed against the inside of the host pipe using a forming tool (Figure 7-5). During the inversion technique, the liner is turned inside out so that the resin is pressed against the inner surface of the host pipe. After heat curing (typically by circulating hot water or steam), the ends of the liner are cut flush with the end of the pipe. To ensure a sealed system at termination points in the lining system, as well as at reinstated service connections in distribution piping, some systems rely on the adherence of the liner resin system to the host pipe. In other cases, lining systems will rely on mechanical fittings to achieve a seal. The system is capable of negotiating a sequence of up to 45° manufactured bends. The range of available diameters is typically 4 to 40 in. (102 to 1,016 mm).

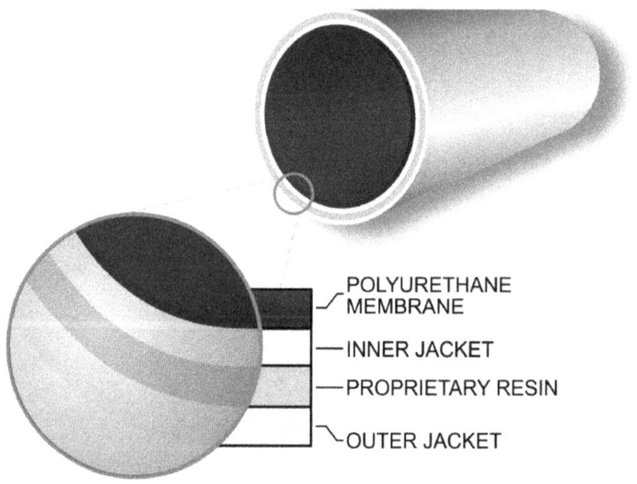


Courtesy of Sanexen Environmental Services, Inc.

Figure 7-4 Cross-section of woven hose

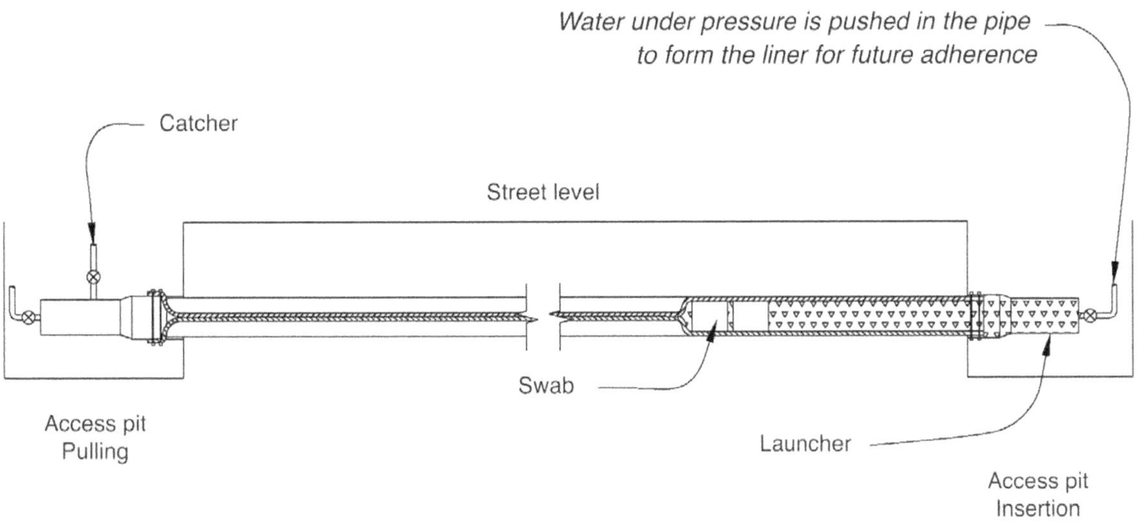


Courtesy of Sanexen Environmental Services, Inc.

**Figure 7-5 Forming of the liner**

Woven hose systems can offer from Class III semistructural capability to Class IV structural capacity depending on the thickness and strength. They can be used in pipes ranging from severe internal corrosion and pinhole leakage, leakage due to faulty joints, and in some cases, problems arising from localized external corrosion to broken or completely deteriorated host pipe situations. The installed liner is very thin, and with its high Hazen–Williams value and joint-free construction, flow rates in the rehabilitated pipeline may become identical to those in the pipeline in its original state. Resistance to external buckling loads can depend on the quality of the adhesive bond to the pipe wall, and consequently, on the quality of cleaning and drying of the host pipe achieved prior to insertion. However, variants of the woven hose system are at least self-supporting (i.e., Class III) when cured, or do not depend on the condition of the host pipe (i.e., Class IV).

Woven hose systems were originally developed in Japan for rehabilitation and earthquake protection of municipal gas lines. The systems are now being used in North America, Japan, and Europe for rehabilitation of potable water pipes.

## Membrane Systems

These systems line pipes with thin elastomeric membranes coated with thermoset resin. Liners are pulled or inverted into the host pipe using air pressure and heat cured in a manner similar to felt-based and woven hose systems. Membrane lining systems were developed for rehabilitation of leaking low-pressure (up to 10 psi [169 kPa]) gas mains. They offer more limited capabilities for spanning holes and gaps at typical water system operating pressures than the previously described systems.

# REFERENCE

American National Standards Institute (ANSI). 2012. Drinking Water System Components—Health Effects. NSF/ANSI 61-2012. 198. Ann Arbor, Mich.: NSF International.

This page intentionally blank.

Chapter 8

# Sliplining

## SLIPLINING AND MODIFIED SLIPLINING

Inserting flexible thermoplastic liners directly into existing water mains can be a viable method of rehabilitation. The liner inserted can be either loose-fitting or close-fitting. This process employed for installing loose-fitting liners, commonly known as *sliplining*, has been widely used by sewer and natural gas utilities since the early 1980s. It is also used in transmission and distribution systems for source water and treated, potable water. Close- or tight-fitting liners are installed using one of a range of variations of sliplining, in which the circular shape of the liner is temporarily modified to reduce the cross-section of the pipe for insertion. These methods involve some form of reversion process to re-round the liner, typically providing a liner to its original or design size. This family of lining techniques is referred to as *modified sliplining*.

Polyethylene pipe is manufactured per ANSI/AWWA Standards C901 (3 in. and smaller) or C906 (4 in. and larger). Polyethylene pipe grade material is classified by ASTM Standard D3350. This material can handle moderately high temperatures (140°F for pressurized lines) and highly corrosive and abrasive liquids in nearly all applications.

Polyvinyl chloride (PVC) pipe is manufactured per ANSI/AWWA Standards C900 (4- to 12-in.) and C905 (14- to 48-in.). PVC pipe grade material is classified by ASTM Standard D1784. Maximum recommended temperature of pipe is 140°F. Fully restrained joints required for pulling long lengths of PVC are accomplished using mechanical spline or pinned connections or through the use of a butt fusion joint.

For some special applications, the use of other thermoplastics such as polypropylene may be recommended.

## SLIPLINING

The key benefit of sliplining is that it creates a new, integral pressure pipe inside the old, deficient pipeline without a complete excavation.

Using a process known as *thermal butt fusion* or *mechanical means of coupling*, the ends of several consecutive lengths of flexible pipe are joined at a convenient location above ground. The individual pipes joined can range from or up to 59 ft (18 m) to form a single length of pipe that can stretch to several thousand feet long. One end of this pipe is then pulled by cable or similar method into the entry pit and through the section of old pipe. The new pipe is then reconnected to the existing mains.

Potential applications for sliplining are numerous. Most existing pipelines can be sliplined, but certain applications are ideally suited to this method:

- Where poor structural integrity of existing pipes makes other lining methods, such as cement mortar, inadvisable
- When service connections and branches are limited
- Where a structure has been built over the existing main, making replacement economically impractical
- Where a main crosses over or under railroads, bridges, rivers, or other obstacles, making alternative linings impractical or not economically feasible
- Where other unique circumstances make alternative lining methods impractical
- Where the lined pipe will provide adequate capacity

An inserted liner does substantially reduce the effective cross-sectional area of the pipe. Consequently, post-lining flow requirements must be considered when deciding to slipline. However, the reduction in the friction factor of the liner pipe as compared to the old, unlined pipe should significantly compensate for the reduced internal diameter. In addition, the flow rate will not be reduced by corrosion, biological growth, or scale over time, as would be expected of many other unlined piping materials. Finally, the geometry of the unlined pipe must be considered, as liners generally do not turn well through elbows.

## Operating Procedures

The type of material for the sliplining pipe should be carefully chosen. To date, the most common material specified is thermoplastic pipe manufactured of high-density polyethylene (HDPE). Publications addressing the details of installing polyethylene pipe as supliners include ASTM Standard F585 and the Plastics Pipe Institute Handbook (2008: 397-420). Recently, PVC pipe has become a more widely used material as well.

To allow for smooth insertion, the insertion pipe should be sized so that the clearance between its outside diameter (OD) and the inside diameter (ID) of the pipe being lined is at least 10 percent of its diameter for pipe sizes up to 20 in. For larger pipe, 2 in. of clearance is generally sufficient. Possible obstructions at pipe joints and taps, and the normal friction created during the insertion process dictate a conservative approach to liner pipe sizing. Pipe sizes are typically standard iron pipe size (IPS) or ductile-iron pipe size (DIPS), but special diameters are also available for sliplining.

Factors to consider in material selection include wall thickness needed to achieve the desired pressure rating at the design temperature and ability to withstand soil loads, external pressure, and traffic loads. A pipe's ability to withstand the rigors of insertion is generally related to the use of proper installation procedures and not the pipe material, because even the thinnest commercially available HDPE or PVC pipes are routinely sliplined in lengths of several hundred feet (100 m or more) to several thousand feet (hundreds of meters).

The pressure rating is determined by the type of piping materials and the standard dimension ratio (SDR or DR), which is the ratio of the outside diameter to the minimum wall thickness. All diameters of pipe having a given SDR value, such as SDR 17 or DR 25, will have the same pressure rating and can be combined into one piping system. Any sliplined pipe is fully pressure rated. Flow through PVC and polyethylene may be calculated using the Hazen–Williams equation as demonstrated in chapter 1. Where negative pressure may occur within the pipe, (1) the pipe must be of a sufficient DR or SDR to resist collapse or (2) grout must be applied to the annular space.

A variety of special fittings, taps, and other appurtenances are manufactured to make transitions between HDPE and PVC pipe and old cast-iron, ductile-iron, and steel mains and to reconnect customers and branches to the newly inserted plastic pipe.

Prior to sliplining a main, a detailed plan should be developed of the existing pipelines, valves, branch pipelines, service connections, and fire hydrants. Because sliplining sometimes requires large excavations, special attention must be given to traffic control and public safety. As the entire existing pipeline is not excavated, any excavations should be located away from highly traveled areas whenever possible. Plans and specifications should clearly spell out required standards of line installation and related work. If the existing pipeline is severely damaged, a cementitious grout may be injected into the annular space between the liner and the existing pipe to provide additional support for traffic and earth loads.

Temporary service lines may be required when lining pipelines for customer service connections or for fire protection. Although downtime required for sliplining can be relatively short, the need for temporary service should be considered.

The designer must decide what work other than actual sliplining will be performed. Directly related to the lining process, of course, is the reconnection of services, fire hydrants, and sideline laterals, and connection of the newly relined system to adjacent sections of the water system. Other work could include replacement of sections and in-line valves, service piping, and corporation cock replacement. Also, fire hydrant and sideline laterals may be sliplined or replaced, and lateral valves and fire hydrants may be replaced.

If temporary service is provided, it should be installed before beginning the insertion process.

A minimum of two excavations are normally required for sliplining, one at each end of the segment of pipe to be sliplined. Commonly, excavations at elbows or other fittings allow equipment to pull in two directions from one location. The length of excavation and amount of host pipe that must be removed at the insertion pit and retrieval pit vary with the type of pipe and depth of installation. Typically at the insertion end, a sloped trench is required to allow liner installation. The trench width should be adequate to allow comfortable operation of pipe-cutting devices and to make reconnections. Extending the length of an excavation reduces the difficulties of insertion, especially in cold weather when the pipe becomes stiffer and more difficult to bend.

The second excavation (the pull hole) should be large enough to accommodate pipe work, pulling cable angles or other pull-in device, and possibly fusion equipment. Lining of large-diameter or deeply buried pipe may require a slightly larger excavation. Excavations should be sheeted and shored as required and bridged with steel plates or securely fenced during nonwork hours for public safety. The distance between excavations varies depending on pipe layout and diameters. Single pulls in excess of 1,000 linear ft (300 m) are possible depending on host pipe alignment and weight of liner. It is recommended that excavations for service and lateral reconnections be made before beginning the pulling procedure.

The section of pipeline to be lined must be isolated by closing valves and then dewatered. (Water may actually make sliplining easier, because PVC and HDPE pipe have nearly neutral buoyancy in water, but its presence may cause difficulty for workers.) A section of existing pipe is then cut out of each end of the section to be lined, leaving a gap long enough to accommodate the pulling process at one end and the liner insertion at the other. Preparation of the carrier pipe may require pipe cleaning, depending on the extent of internal corrosion and tuberculation buildup. After cleaning, video inspection of the host pipe should be considered to ensure that no unforeseen or expected conditions exist, such as abandoned valves or unknown service connections, etc. In most cases, sections of the old pipe must be cut out to facilitate reconnection of services and other laterals.

Once the carrier pipe is properly prepared, a winch cable is fed through it and attached to a bullet-shaped pulling head on the HDPE or PVC liner. The liner is constructed aboveground at a convenient location by joining sections of HDPE or PVC pipe with a butt-fusion machine. This machine produces a true joint with the same structural integrity and tensile strength as the pipe itself. When laying down the pipe for pull-in, surface obstacles pipe may be avoided by stringing the pipe along a curved path. Pipe manufacturers publish the minimum radius of curvature. HDPE ranges from 20 to 40 times the pipe diameter depending on the SDR. (For example, 27 × OD for SDR 17 pipe and 25 × OD for SDR 11 pipe.) An example of the slipline insertion process is illustrated in Figure 8-1. Figure 8-2 shows a typical exit of the slipline pipe material and pulling head.

Small diameter HDPE liner pipe (up to 6.625-in. [168-mm] OD) can be purchased in coils to reduce the number of field-fused joints; however, the added friction caused by pulling pipe that has attained a set curve during coiling must be considered. The liner is then pulled through the carrier pipe with a power winch or other pulling device. When installing large-diameter liners, a backhoe or bulldozer is often used in a push-pull technique, in which equipment pushing the liner assists the pulling action of the winch. Caution must be exercised during insertion to prevent gouging the exterior of the liner as it is pulled into the existing pipe. Flexible liners include a 10 percent wall thickness allowance for gouging or scratching.

After insertion is completed, the liner must be connected to the remaining water mains. For HDPE applications, increasers fused to the end of the new liner with suitable mechanical fittings, or mechanical reducer fittings can be used join the liner to the existing pipe. For HDPE liners, flanged plastic fittings, known as *flange adapters*, with metal backup rings can be used to terminate the lined section and reconnect to the existing system. Mechanical joint adapters that meet ANSI/AWWA C111/A21.11-12 are available to connect lined segments or to bolt the new lining to the existing steel, cast iron or ductile iron. For PVC, mechanical adapters in both flanged and mechanical joint configurations are available to connect the PVC pipe to a valve or fitting. Standard reducers and increasers are used to connect back to the original host pipe.

Service lines, fire hydrants, and branch laterals that have been excavated and cut out of the carrier pipe must be connected to the liner. For HDPE, service corporations are thermally saddle-fused directly to the liner and tapped with a self-tapping tee. For PVC, standard tap saddles and sleeves are used to make service connections, following standard PVC tapping procedures. Workers may have to cut in tees and make connections to existing branches for fire hydrant or sideline laterals, similar to those previously described for main line reconnections. Strict attention must be paid to the pressure ratings of all fittings. Figure 8-3 shows the reconnection of the liner to the existing distribution system.

The inserted and reconnected liner should be tested just as any newly constructed water main would be. Test pressure should not exceed 150 percent of the pressure rating of the pipe. For HDPE, the test pressure should not be applied for more than 8 hr without a period of relaxation in between. HDPE pipe will expand in diameter during this type of testing, requiring regular additions of water to maintain the test pressure. ASTM Standard F2164, *Field Leak Testing of Polyethylene Pressure Piping Systems Using Hydrostatic Pressure* should be used for testing guidelines. Fused PVC is tested according to ANSI/AWWA Standard C905 guidelines.

Courtesy of Underground Solutions, Inc.

Figure 8-1 Insertion of fused PVC for a 3,500 ft slipline

Courtesy of Underground Solutions, Inc.

Figure 8-2 Exit of the sliplined pipe from host pipe

Courtesy of Underground Solutions, Inc.

**Figure 8-3 Liner prepared for reconnection**

Special attention must be given to exposed sections of the liner at pull or insertion excavations and at lateral reconnection points. Proper bedding and backfilling are necessary to prevent differential settlement.

Final completion of the lining process consists of restoring the main to service following disinfection, removing temporary bypass service, permanent paving, and site cleanup.

Drawbacks to the sliplining method include a reduced cross-sectional area and the numerous excavations that must be made if many service and branch reconnections are involved. Regardless of these drawbacks, sliplining is a useful lining method for the water utility engineer or manager.

## MODIFIED SLIPLINING TECHNIQUES

Modified or close-fit sliplining techniques involve inserting a thermoplastic tube, which has had its cross-section temporarily modified to allow sufficient clearance for insertion, into the host pipe. When the section-modified tube is subsequently returned to its approximate original shape, it provides a close- or tight-fit lining in the host pipe. Depending on the process selected, the cross-section modification can be made in the factory or on-site.

Two key differences distinguish this class of techniques from conventional sliplining.

1. Greater retention of hydraulic cross-section. Thin liners may actually improve the flow capacity of the existing pipe, despite the reduction in cross-sectional area, through improved smoothness (increased Hazen–Williams C factor) and lack of joints in the liner.
2. Flexibility of liner thickness. Liner thickness can be selected to provide either fully structural (Class IV) or semistructural (Class III) internal pressure capability. The latter option can provide a more cost-effective system of renovation than other options for some types of host pipe deterioration.

## Classification of Systems

Modified sliplining systems can be classified into three main groups.

## Symmetrical Reduction Systems

These systems involve the use of a round thermoplastic tube with an OD the same as or slightly larger than the ID of the host pipe. The tube is passed through either a static die or an array of compression rollers, which temporarily reduce its diameter to allow sufficient clearance for insertion into the host pipe. After winch insertion, the tube is allowed to revert toward its original dimensions. In some applications, this reversion is facilitated by the application of internal pressure. These techniques generally use polyethylene pipe to take advantage of the material's molecular "memory" for the dimensions formed at the time of extrusion.

## Folded and Formed Systems

These lining systems involve modification of the cross-section of a thermoplastic tube into a folded C- or U-shape, either at the manufacturing plant or in the field. After winch insertion into the host pipe, the tube is returned to its original shape and diameter using heat and/or pressure. These techniques can be applied to polyethylene, fiber reinforced polyethylene, and PVC liner tubes.

## Expanded PVC

This system uses a PVC pipe that is inserted by sliplining into a prepared host pipe. The PVC pipe formulation has been developed for the expansion process and meets all required standards for ANSI/AWWA Standards C900 and C905 and ANSI/NSF Standard 61. Mobile equipment heats the PVC pipe and expands it into the host to a close-fit liner. The expanded PVC is cooled in place retaining its new dimensions.

Table 8-1 presents a summary of modified sliplining methods available for HDPE pipe with a brief description of the processes.

**Table 8-1 HDPE modified sliplining methods**

| Modification Type | Modification Location | Modification Device | Modified Shape | Reversion Method | Examples |
|---|---|---|---|---|---|
| Symmetrical tension | On-site | Static die | Circular | Natural | Swagelining |
| Symmetrical tension | On-site | Roller die | Circular | Natural | Tite liner |
| Symmetrical compression | On-site | Multiple rollers | Circular | Water pressure | Insituguard Flex, Rolldown |
| Fold and form | Factory | Hot fold | "U" or "C" | Steam and pressure | U-liner, Compact pipe |
| Fold and form | Factory | Cold fold | Heart or celery | Air/water pressure | Subcoil |
| Fold and form | On-site | Cold fold | Heart or celery | Water pressure | Insituguard Flex |
| Fold and form | On-site | Cold fold | Heart or celery | Air/water pressure | Subline |

## SYMMETRICAL REDUCTION SYSTEMS

A number of commercially available systems provide a wide range of solutions for inserting liners. These systems differ in the methods that are used to achieve the reduction in diameter, the methods and time scales for reversion to the design shapes, and the insertion onto the host pipe. The class of liner that can be provided also varies with the system implemented.

### Static Die Systems

Two such systems were originally developed in the United Kingdom by major gas and water utilities in response to the need for a trenchless renovation technique for municipal gas and water mains. Both involve pulling a length of polyethylene pipe through a static-diameter reduction die directly into the pipe to be renovated and maintaining winch tension until the pipe is fully inserted. When the winch tension is released, the polyethylene pipe rapidly reverts to its original diameter, simultaneously shrinking in length, until it achieves a close fit in the host pipe. The original process used an oven to heat the pipe before it passed through the die in order to reduce the winch forces required. Currently, this process is rarely used.

These processes have been mainly applied to cast-iron, ductile-iron, and steel gas and water mains in the diameter range 4 to 18 in. (102 to 457 mm). One process has been used, however, in water transmission pipe up to 42 in. (1,067 mm) in diameter.

### Roller-Based Systems

Initially, two roller-based processes were developed to meet the needs of the US oil, gas, and mining industries for a thin-wall polyethylene lining system to control internal corrosion in steel pipes carrying oil, gas, and aqueous products at high pressures. Both procedures involve pulling a length of polyethylene pipe through a series of reduction rollers, which may be hydraulically driven and/or pushed, directly into the pipe to be renovated. The processes involve somewhat lower winch tensions than static die techniques require, and the liner may not revert toward original dimensions quite as rapidly when winch tension is released. These processes have been used for insertion of long (2,300-ft [700-m]) sections of pipe in diameters up to 53 in. (1,346 mm). Most field experience has involved lining welded steel pipelines, although some installations have inserted linings in cast-iron and ductile-iron municipal gas and water lines.

A third roller-based process, developed in the United Kingdom in parallel with the die-based processes, has been extensively used for renovation of cast-iron gas mains, and subsequently water mains, in the diameter range of 4 to 18 in. (102 to 457 mm). The process involves pushing a length of polyethylene pipe through a series of reduction rollers. In contrast to the static die systems and the other roller based systems, a large part of the reduction in diameter is retained for a period ranging from hours to days, depending on ambient temperature. Full reversion to final dimensions requires the application of a high internal water pressure for 12 to 24 hr. This time lag allows insertion of the reduced pipe into the host either directly, as with the other processes, or at a different time and location.

Recently, another roller-based system has been introduced for application in water main renovation for liners ranging from 6 to 24 in (150 to 600 mm). Like previous systems, it utilizes a series of reduction rollers; however, each roller in all roller panels (typically three to four panels) is hydraulically driven. A large component of the diameter reduction is maintained for a period ranging from hours to days, depending on ambient temperature, and full reversion to design dimensions requires iterative application of high internal pressure (typically two to three times the design pipe operating pressure). In this system, the power rollers can provide a significant portion of the driving force to insert the liner, and the winch force is significantly reduced, generally providing a guiding tension. This system is designed to minimize the longitudinal stress introduced into the liner pipe, reducing the potential for retraction if the liner has to be cut in the future. This situation may occur during installation of a new fitting or for repair if damaged by external causes. Figure 8-4 shows an example of a roller-based symmetrical reduction liner machine. Figure 8-5 shows the fusion of an expansion head on the pipeline for the symmetrical reduction process.

## General Characteristics

All of the symmetrical reduction techniques share the following characteristics.

The energy required to reduce the polyethylene pipe diameter increases dramatically with pipe size and wall thickness (indicated by SDR). Therefore, winch tensions and/or hydraulic power requirements may rise with increasing diameter and thickness, limiting the maximum pipe wall thickness that can be handled at each diameter. These limitations vary from process to process and can affect the ability of the processes to be used for renovating thick, Class IV polyethylene liners. Depending on the system used, there may be sensitivity to time to insert after reduction.

Care is needed in lining pipes with significant local variations in internal diameter because of manufacturing tolerances or other potential obstructions such as joint offsets. These variations may reduce or even eliminate insertion clearance at certain points.

Courtesy of Insituform Technologies, LLC

Figure 8-4 Roller-based symmetrical reduction liner machine

Courtesy of Insituform Technologies, LLC

Figure 8-5 Fusing expansion head for symmetrical reduction process

The required polyethylene pipe diameters are rarely standard pipe sizes. Manufacturing pipes in needed sizes may involve the use of special extrusion dies, with potential implications for the cost and timing of projects. The processes are sometimes unable to negotiate manufactured bends, and local excavation is required at the locations of these service connections and any other in-line fittings. Typically, some form of lubrication, such as bentonite or other liquid product, is required to reduce friction during installation.

Sufficient site space is required to accommodate butt-fusion welding of polyethylene pipes into long sections the full length of the lining segment prior to diameter reduction and during subsequent insertion. For the processes that require water pressure for reversion, the ends must be capped with fittings capable of withstanding pressures up to three times the rated operating pressure of the liner.

## FOLDED AND FORMED SYSTEMS

Factory folded/hot rerounded systems involve tubing heated and folded into a C- or U-shape in the factory and then transported to the job site on a reel that may contain up to 2,000 ft (610 m) of liner, depending on diameter. The folded liner is winched into the host pipe (in some cases, after reheating the liner on the reel to facilitate placement). Then it is rerounded using a combination of heat (usually provided by steam) and pressure (provided by steam and/or air). The rerounding process may be progressive, using a device that is propelled through the liner, or it may occur simultaneously throughout the pipe length.

Two such systems use polyethylene pipe to line mains ranging from 4 to 16 in. (102 to 406 mm) in diameter with liner pipe thickness in the range SDR 21.0 to 32.5. In a recent development, large-diameter (21-in. [533-mm]) folded polyethylene pipes have been shipped to job sites in 40-ft (12-m) lengths, where factory-matched lengths are joined by butt-fusion welding into long (1,500-ft [457-m]) sections prior to insertion and rerounding. This technique allows use of larger diameters than are possible with the more conventional reels of folded material. Such systems have been widely used for wastewater pipes, and they have been evaluated for gas and potable water lines.

In addition, at least two sliplining systems use modified rigid PVC, mainly for wastewater pipelines. The manufacturers of these systems are currently exploring their potential for use in water pipelines.

Polyester-reinforced polyethylene (PRP) liners consist of a circular woven polyester tube, encapsulated in polyethylene on the internal and external faces. This system was developed in the United Kingdom specifically for water main renovation and has been used extensively since the mid-1990s. It is currently used throughout Europe, North America, Asia, and other areas. Originally available in 3- to 6-in. (76- to 152-mm) sizes, the size envelope has been expanded to 12 in. (305 mm). The reel-wound PRP liner is delivered to the site in folded form and then winched at ambient temperature into the host pipe. The liner is then rapidly rerounded using a combination of air and steam, under low pressure. The process results in a structural Class IV liner, with an internal pressure capability of 170 psi (1,170 kPa). In the 4- through 8-in. sizes, a liner capable of internal pressure of 230 psi (1,585 kPa) can be provided. Wall thickness can range from 0.08 to 0.20 in. (2 to 5 mm), depending on diameter and pressure rating. The liner can negotiate manufactured bends of up to 45° with some internal wrinkling.

Site cold-folded/cold rerounded systems involve passing lengths of thin, round polyethylene pipe through a piece of site-based equipment that folds the pipe into a U-shape. The shape is restrained by a series of thin plastic straps applied to the folded pipe as it exits from the folding machine. After winching into the host pipe, the polyethylene liner pipe is rerounded through the application of internal pressure, which breaks the straps.

This method is generally applicable to pipe thicknesses up to SDR 26 or DR 21, governed by wall thickness. It has been used in Europe as a Class III liner for water pipes up to 36-in. (914-mm) diameter and in New York for a 48-in. (1,200-mm) diameter water main. The envelope is expanding to include Class IV liners, and sizes to 60 in. (1,500 mm). It offers the advantage over factory-folded techniques that additional lengths of polyethylene pipe can be fused to the lining section at any time prior to passage through the folding equipment and during insertion of the liner into the host pipe. Hence, the method can accommodate long insertions at large diameters with minimal site space problems. The polyethylene pipe is normally of nonstandard size produced with OD equal to the minimum expected ID of the host pipe. Figure 8-6 shows an example of a site-folded HDPE liner.

## General Characteristics

All of the folded and formed techniques share the following characteristics.

- They involve much larger insertion clearances and are not as time sensitive as most symmetrical reduction techniques; hence they are much less sensitive to local variations in pipe diameter.
- The folded shape generally allows insertion through a smaller entry pit than the symmetrical reduction techniques require.
- Typically some form of lubrication, such as bentonite or other liquid product, is required to reduce friction during installation.
- Where such systems are installed as Class IV liners, confirm that the folding and rerounding process has not affected the long-term pressure capability of the liner.
- Reversion, or rerounding, requires that the ends of the liner are capped with suitable fittings.

Courtesy of Insituform Technologies, LLC

Figure 8-6 Site-folded HDPE liner

## EXPANDED PVC SYSTEMS

These systems involve expanding specific PVC pipe and increasing its diameter permanently. PVC pipe is typically used on host pipe nominal sizes from 4 to 12 in. PVC pipe specifically formulated for fusion and expansion is butt fused together in lengths of up to 500 ft. This resulting single pipe length is then inserted into the host pipe via sliplining.

After insertion, end hardware is assembled to allow the circulation of hot water through the pipe and to determine the end configuration of the expanded pipe for fittings and reconnection. The pipe ends can be expanded to the nearest IPS or DIPS size for use of standard connection hardware for reinstatement.

When the pipe is heated to the appropriate parameters, it is expanded to meet the ID of the host pipe and the end connections. While under pressure, the pipe is cooled in place holding the expanded dimensions. Upon completion, the end hardware is removed, the expanded pipe trimmed to length, and the reconnections made.

The selection of the starting stock wall thickness and the amount of expansion to be done allows the liner to be sized for most normal water system pressure applications in the 150 psi operating pressure range. The liner is normally pressure tested to 1.5 times the operating pressure of the system before reconnection. The liner does not require nor does it use any contribution from the host pipe for its pressure rating, making it a true Class IV fully structural lining.

## LINER TERMINATION FITTINGS

An integral component of any lining system is the end fittings. These are required for reconnection of a lined segment to the existing system, or two lined segments. In some cases these fittings eliminate the potential for the liner to retract after it is installed. A wide range of options is available, and a suitable fitting must be selected in the design of a lining system (Figures 8-7, 8-8, and 8-9).

Courtesy of Insituform Technologies, LLC

Figure 8-7 Special liner grip fitting

Courtesy of Insituform Technologies, LLC

Figure 8-8 Special liner grip fitting

Courtesy of Insituform Technologies, LLC

Figure 8-9 Special multi-grip fitting

End fittings can be fused-on, mechanical, or push-on, depending on the liner design and application. Some mechanical compression-type fittings require internal stiffeners to prevent the collapse of the liner as they are tightened. Some mechanical fittings can provide an end restrained solution. Fused-on fittings can be used for structural liners, because the liner must protrude from the host pipe for connection. These can be either butt or electrofusion.

## REFERENCES

American National Standards Institute (ANSI). 2012. Drinking Water System Components—Health Effects. NSF/ANSI 61-2012. 198. Ann Arbor, Mich.: NSF International.

American Society for Testing and Materials (ASTM). 2010. Standard Practice for Field Leak Testing of Polyethylene (PE) Pressure Piping Systems Using Hydrostatic Pressure. ASTM F2164-10e1. Developed by Subcommittee: F17.40 Standards Volume: 08.04. West Conshohocken, Pa.: ASTM.

ASTM. 2011. Standard Specification for Rigid Poly (Vinyl Chloride) (PVC) Compounds and Chlorinated Poly (Vinyl Chloride) (CPVC) Compounds. ASTM D1784–11. Developed by Subcommittee: D20.15 Volume: 08.01. West Conshohocken, Pa.: ASTM.

ASTM. 2012. Standard Specification for Polyethylene Plastics Pipe and Fittings Materials Active Standard. ASTM D3350–12. Developed by Subcommittee: D20.15 Volume: 08.02. West Conshohocken, Pa.: ASTM.

ASTM. 2013. *Standard Guide for Insertion of Flexible Polyethylene Pipe into Exisiting Sewers*. F585. West Conshokoken, Pa.: ASTM.

American Water Works Association (AWWA). 2006. Standard for Rubber-Gasket Joints for Ductile-Iron Pressure Pipe and Fittings. ANSI/AWWA C111/A21.11. Denver, Colo.: AWWA.

AWWA. 2007. Standard for Polyvinyl Chloride (PVC) Pressure Pipe and Fabricated Fittings, 4 In. Through 12 In. (100 mm Through 300 mm), for Water Transmission and Distribution. ANSI/AWWA C900. Denver, Colo.: AWWA.

AWWA. 2008. Standard for Polyethylene (PE) Pressure Pipe and Tubing, ½ In. (13 mm) Through 3 In. (76 mm), for Water Service. ANSI/AWWA C901. Denver, Colo.: AWWA.

AWWA. 2010. Standard for Polyvinyl Chloride (PVC) Pressure Pipe and Fabricated Fittings, 14 In. Through 48 In. (350 mm Through 1,200 mm). ANSI/AWWA C905. Denver, Colo.: AWWA.

AWWA. 2012. Standard for Polyethylene (PE) Pressure Pipe and Fittings, 4 In. Through 65 In. (100 mm Through 1,600 mm). ANSI/AWWA C906. Denver, Colo.: AWWA.

Plastics Pipe Institute. Pipeline Rehabilitation by Sliplining with Polyethylene Pipe. *The Plastics Pipe Institute Handbook of Polyethylene Pipe*. 620:2008.

This page intentionally blank.

Chapter 9

# Internal Joint Seals

Internal joint sealing is a method of in-situ pipe joint rehabilitation that is used to correct leaking pipe joints, to strengthen weak joints, repair pipe cracks causing leaks, prevent infiltration, or in concert with other nonjoint specific methods of rehabilitation (cement–mortar lining, cured-in-place pipe lining, etc.). The installation of the seals requires manned entrance into the pipe and is therefore generally limited to pipes sizes of 16 in. in diameter or greater. Seals are typically manufactured to withstand internal working pressures up to 300 psi though seal for higher pressures are also available. The distance between street access holes can range up to 5,000 feet. Seals are generally available in a variety of widths. The standard width is used on joint gaps up to 5-in. wide whereas extra-wide widths are applicable for joints up to 8 in. in width and the double-wide are applicable for joints or gaps up to 13 in. Additionally, custom seals can be designed for larger widths via the use of rubber sleeves.

Internal joint seals incorporate a number of lip seals on their outer edges that together completely seal the circumference of the pipe on either side of the joint. The seal's flexibility ensures a bottle-tight seal around the entire pipe joint, while its low profile and graded edge allows water to flow without creating turbulence.

Internal joint seals for potable water applications are manufactured from an ethylene propylene diene monomer (EPDM) synthetic rubber compound. (Nitrile rubber is available for use in nonpotable water applications). All materials used in internal joint sealing of potable water mains should meet the requirements of ANSI/NSF Standard 61 for use in potable water systems. See Table 9-1 for additional details on materials used in internal joint seals.

Seals are packed for shipping in a manner that will not damage or deform them, and special precautions must be maintained until the seal is fitted within the main. Seals must be stored in a dry environment at room temperature, and they should not be allowed to remain in direct sunlight.

Prior to fitting, seals should undergo a thorough visual examination by operators or certified installers, paying particular attention to the ribbed (lip seals) sections of the seal. If there is doubt, a seal should not be used.

**Table 9-1 Material details for internal joint seals***

| Type of Joint Seal | Material |
|---|---|
| Retainer bands | Type 304 stainless steel, 3/16-in. (4.8-mm) thickness × 2-in. (51-mm) wide strips; ultimate tensile (85,000) yield strength (35,000) meets ASTM Standards A-167, A-267, and A-479-85 |
| Shims | Type 304 stainless steel, 0.048-in. (1.22-mm) thickness × 2-in. (51-mm) wide × 6-in. (152-mm) long |
| Test valve, low profile | Type 304 stainless steel |
| Plastic backing band | Material meets ASTM specification D 1248 Sub Type 3. This material is a high-density polyethylene extrusion with a specific gravity of 0.960. |
| Joint mounting surface preparation | Nontoxic thread seal compound and/or nontoxic vegetable lubricant |
| Joint gap filler | Portland cement mortar, Type 5 meets ASTM Standard C150 epoxy for underwater repair and grouting |
| Pipe surface gel | 100% solid epoxy nontoxic patching gel or equivalent |
| Internal joint seal | Certified to ANSI/NSF Standard 61 EPDM rubber for potable water |

* Internal joint seals meet ASTM specification D2000.

## FITTING PROCEDURE FOR INTERNAL JOINT SEALS

All procedures covering safety of personnel working within a pipeline must be carried out in accordance with applicable safety regulations.

1. **Joint preparation.** Experience has shown that most deleterious deposits can be removed from pipe walls around a joint by hand scraping and brushing. Occasionally, power tools are used to remove stubborn deposits or hard-scale lamination. Whichever method is adopted, the main must be as clean as reasonably possible to provide an acceptable working environment for the operators. The joints themselves must be completely clean.
2. **Joint filling.** The joints are filled with Portland cement to the full depth of the gap and rendered flush with the internal surface of the pipe. All surplus material and spillage should be removed from the joint area prior to the surface preparation for the seal. The joint filling operation should always be carried out before final preparation.
3. **Surface preparation of the joint area.** The area of pipe on either side of the joint where the actual lip seals make contact must be prepared. The resulting finish must allow the lip seals to bed consistently so they provide a permanent seal (Figure 9-1).
   a. All high and low spots that form surface imperfections running axially through or part way through the sealing surface must be removed. Deep imperfections must be filled with approved compounds. The material must be smoothed to match the surface of the joint area.
   b. Circumferential grind marks are allowed, provided they do not exceed 0.030 in. (0.76 mm) in depth.

c. The extent of the prepared area on either side of the joint must be compatible with the lip seals, and the prepared area must extend at least 1 in. (25 mm) beyond either side of the ribbed section of the seal.

d. The pipe should be marked with grease chalk to define the prepared areas and seal position.

Note: The importance of good surface preparation cannot be overemphasized.

4. **Surface lubrication.** Immediately prior to fitting the seal, the area must be cleaned with a dry brush and coated with a nontoxic lubricating soap compatible with the composition of the internal joint seal. The lubricant is hand applied over the entire prepared area using an ordinary paintbrush. Care must be taken not to pick up dust deposits from the unprepared surface and deposit them into the lubricant. The lubricant is purely an aid to fitting the seal and in no way contributes to its sealing capabilities. Before using a lubricant product, confirm its acceptability for use in contact with potable water (Figure 9-2).

5. **Positioning the seal.** Confirm that the seal is undamaged and the test unit (stem and back nut) is tightened to a torque of 15 to 17 in.-lb (1.7 to 1.9 N-m) or in accordance with the manufacturer's recommendations. The internal joint seal is placed in position bridging the joint gap, guided by the previously drawn chalk marks. The seal must be positioned accurately on the prepared areas. The test unit in the seal must be located at either the 9 or 3 o'clock position or in accordance with the manufacturer's recommendations.

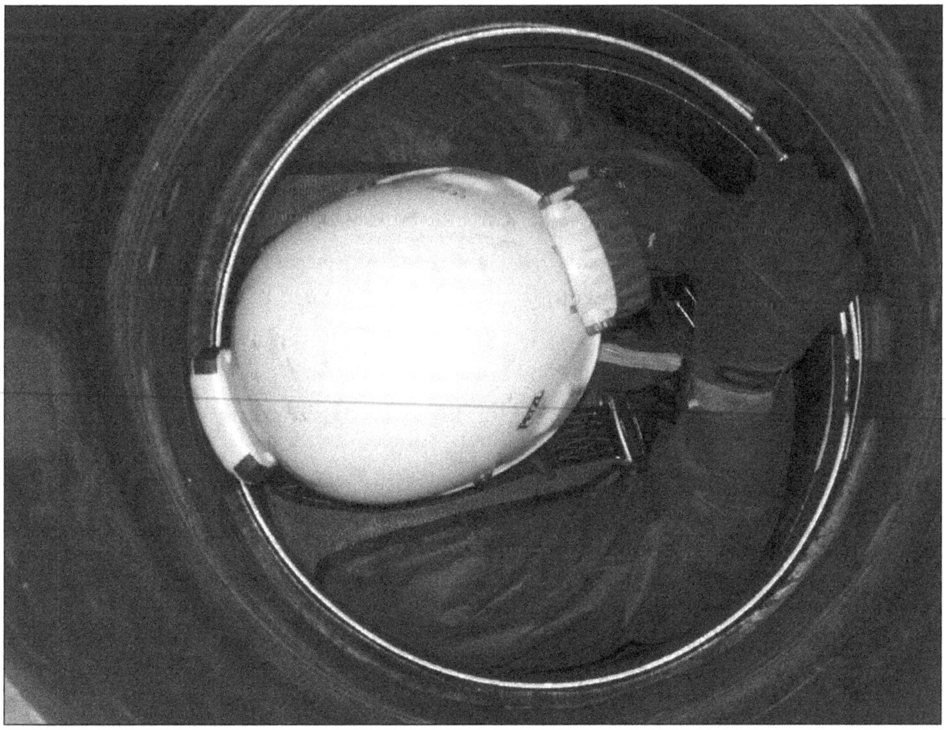

Courtesy of Hydratech Engineered Products, LLC.

Figure 9-1 Joint area is cleaned and prepared prior to installation of the seal

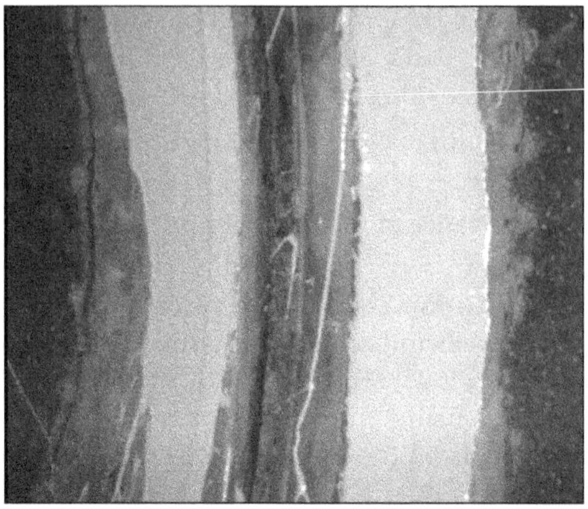

Courtesy of Miller Pipeline, Inc.

Figure 9-2 A nontoxic lubricating soap is applied, and the seal is carefully positioned with its retaining band

The test unit mentioned is generally referred to as a *valve*. This assembly has no integral sealing capability. It is sealed only by the fitting of the plug. Position the backing band (if required) behind the seal over the joint area.

This component is sometimes easier to install after one of the retaining bands has been set in place.

6. **Positioning retaining bands.** Before stainless steel retaining bands are placed in the grooves provided in the seal, two 0.048-in. (1.22-mm) stainless steel radial shims are placed in these grooves at the band gap to provide a bridge that will continue the radial load transmitted to the internal joint seal when the bands are expanded. Both bands are temporarily locked in position, with their ends equally spaced across the spring steel shims. Band dimensions and material vary depending on diameter and pipeline product, but a common unit is 0.1875-in. × 2-in. (4.8-mm × 51-mm) 304 or 316 stainless steel.

7. **Expanding the seal into position.** A hydraulic expanding device called a *ring expander* is used to apply the correct pressure to the retaining bands.

   a. When positioning the expander in line with a retaining band, ensure that the band remains in the groove of the seal and does not move or become dislodged. Also ensure that the expander is positioned correctly in both planes. For example, should the expander be placed by touching the bands at the invert, the expander can be locked at full pressure without exerting any load at all at the top of the seal.

   b. The expander is expanded radially with a predetermined pressure (not more than 4,500 psi [31,027 kPa] on the pump gauge) transmitting the required load against the retaining band and the internal joint seal.

   c. Table 9-2 gives band pressures for various seal sizes. This pressure is held for at least 2 min.

   d. A space provided in the expander exposes the cleats of the retaining band. A locking piece called a *wedge* is fitted between the exposed gap of the expanded band ends. The size of the wedge is selected that gives a slight interference fit between the band ends. The wedge is tapped leading edge first into position, locking in the compression of the internal joint seal. The radius of the wedge is matched to suit the pipe diameter.

e. The pressure is released from the expander and the procedure is repeated on the second retaining band of the seal.

f. This entire re-expansion operation must be completed after at least 1 hr has elapsed after the first expansion. This precaution allows for any seal relaxation that may take place, in which case a slightly wider wedge can usually be fitted. The load forces transmitted by the expander have been determined from test data and should not be altered by changing the pressure over the center of the seal (Figure 9-3).

8. **Extra-wide and double-wide seals.** Fitting procedures are identical to those of standard seals. Sometimes additional retaining bands (internal bands) are placed in the center of the seal to provide additional support. This is a function of the depth of the line and the potential for infiltration.

9. **Testing the seal—Test 1.** Two individual pressure tests are applied to each seal. In the first test, air is introduced at a pressure of 5 psi (34 kPa) into the seal through the valve in the internal joint seal. This pressure is sustained while soapy water is applied to the outer edge and entire body of the seal. Any leak, indicated by a growing bubble or stream of bubbles, must be stopped.

10. **Testing the seal—Test 2.** The second test is applied after each section has been completed and the seal has had time to set. Air is introduced at a pressure of 10 psi (69 kPa), but a restraining device is fitted over the center of the seal to prevent excessive ballooning to the membrane that will occur at this higher pressure. Soapy water is again used to detect any leaks, which must be sealed.

**Table 9-2 Retainer band expansion pressures**

| Diameter | Pneumatic Expander | Hydraulic Expander |
|---|---|---|
| 16 in. (406 mm) | 400 psi (2,758 kPa) | 2,700 psi (18,616 kPa) |
| 18 in. (457 mm) | 400 psi (2,758 kPa) | 2,700 psi (18,616 kPa) |
| 20 in. (508 mm) | 400 psi (2,758 kPa) | 2,700 psi (18,616 kPa) |
| 24 in. (610 mm) | 400 psi (2,758 kPa) | 2,700 psi (18,616 kPa) |
| 30 in. (762 mm) | 400 psi (2,758 kPa) | 3,800 psi (26,200 kPa) |
| 36 in. (914 mm) | 400 psi (2,758 kPa) | 3,800 psi (26,200 kPa) |
| 42 in. (1,067 mm) | 400 psi (2,758 kPa) | 3,800 psi (26,200 kPa) |
| 48 in. (1,219 mm) | 400 psi (2,758 kPa) | 4,000 psi (27,579 kPa) |

Notes: 16-in. (406-mm) through 48-in. (1,219-mm) bands are one-piece units.

Caution should be used during hydraulic expanding of retainer bands; hydraulic expansion of the retainer bands requires a slower rate of expansion due to cold flow of seal.

Apply expansion pressure at a slow rate until maximum allowed gauge pressure is reached. Rapid expansion will result in damage to retaining bands.

54-in. (1,372-mm) through 144-in. (3,658-mm) retaining bands expand at 4,000 psi (27,579 kPa) for two- and three-piece bands. Apply pressure at all wedge points.

In certain pipe materials and pipe-wall thicknesses, the expansion pressure may be adjusted, based on the manufacturer's recommendations, to properly bed the lip seals.

11. **Testing an extra-wide or double-wide seal.** Testing procedures similar to those for standard seals should be carried out. Care must be exercised, however, to ensure that the test pressure does not exceed 5 psi (34 kPa) if the test is conducted without the central restraining device, in order to check the body of the seal. At higher pressures, excessive ballooning will occur resulting in movement of the sealing faces. On seals with very large diameters and where multisection retaining bands have been used, it is not possible, or desirable, to conduct tests above 5 psi (34 kPa) because no central restraining device is available (Figure 9-4).
12. **Completed seal and completion report.** The completed seal provides a permanent solution to leaking joint problems. Finally, a completion report should be supplied detailing the type and location of each seal and any other repair/rehabilitation activities.

Courtesy of Miller Pipeline, Inc.

Figure 9-3 An expansion ring is placed over each retaining band and a wedge is inserted between the band ends

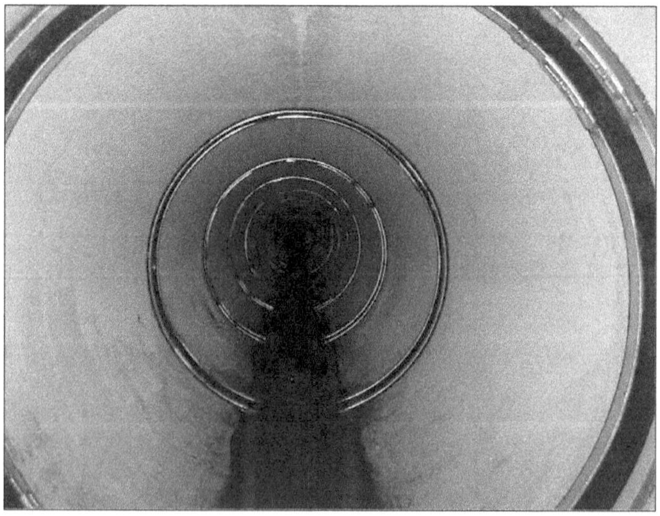

Courtesy of Hydratech Engineered Products, LLC

Figure 9-4 Completed section showing installed internal joint seals

## REFERENCES

American National Standards Institute (ANSI). 2012. Drinking Water System Components—Health Effects. NSF/ANSI 61-2012. 198. Ann Arbor, Mich.: NSF International.

American Society for Testing and Materials (ASTM). 1999. Standard Specification for Stainless and Heat-Resisting Chromium–Nickel Steel Plate, Sheet, and Strip. ASTM A167–99. Developed by Subcommittee: A01.17 Volume: 01.03. West Conshohocken, Pa.: ASTM.

ASTM. 1954. Specification for Lead Coating (Hot-Dip) on Iron or Steel Hardware. ASTM A267 (Withdrawn, no replacement.) West Conshohocken, Pa.: ASTM.

ASTM. 2011. Standard Specification for Stainless Steel Bars and Shapes for Use in Boilers and Other Pressure Vessels. ASTM A479/A479M–11. Developed by Subcommittee: A01.17 Volume: 01.03. West Conshohocken, Pa.: ASTM.

ASTM. 2012. Standard Classification System for Rubber Products in Automotive Applications. ASTM D2000–12. Developed by Subcommittee: D11.30 Book of Standards Volume: 09.02. West Conshohocken, Pa.: ASTM.

ASTM. 2012. Standard Specification for Polyethylene Plastics Extrusion Materials for Wire and Cable. ASTM D1248–12. Developed by Subcommittee: D20.15 Volume: 08.01. West Conshohocken, Pa.: ASTM.

ASTM. 2012. Standard Specification for Portland Cement. ASTM C150/C150M–12. Developed by Subcommittee: C01.10 Volume: 04.01. West Conshohocken, Pa.: ASTM.

# Chapter 10

# Pipe Bursting

## HISTORY

Pipe bursting is a well-established trenchless method of pipeline rehabilitation or replacement that has been in use in the United States for several decades. The process was first developed in the United Kingdom in the late 1970s as a process for replacing natural gas pipelines without having to excavate a trench for the entire existing alignment. The pipe bursting process is a form of pipeline rehabilitation that replaces the original pipe with a new pipe that is the same diameter or larger. Once the process started being performed in the United States (U.S.), the technology was also modified and used to replace water and wastewater pipelines. Existing water mains between 4 in. (102 mm) and 36 in. (914 mm) in diameter have been successfully pipe burst in the United States. More than 20 million ft (over 6,000 km) of pipe has been replaced using pipe bursting since 1994. This chapter focuses the pipe bursting process of existing water mains.

## PROCESS OVERVIEW

The pipe bursting process consists of either a pneumatic (air powered) or static (hydraulically powered) pipe bursting system that fractures or splits the existing water main while simultaneously expanding the surrounding soils and installing a replacement pipe of equal or larger diameter and pressure rating. The pneumatic pipe bursting tool is guided and pulled through the existing water main with a constant tension variable speed winch. The size of the winch being used is dependent on the pipe sizes being pipe burst, length of runs, and soil conditions. The winch is used to guide and provide positive constant tension to move the tool and replacement water main forward. A static pipe bursting system uses rods that are pushed into the existing water main then attached to *roller cutters* (as needed depending on the host pipe material), the expander, and new pipe that is pulled in during the process. The pipe bursting tooling is designed to utilize the existing pipeline as a guiding alignment for the replacement pipe. The bursting tooling moves through the existing pipe while fragmenting or splitting the existing pipe and compressing the original pipe materials into the surrounding soil as the tool and replacement pipe progress. The expander, which is attached to the new pipe being pulled in, is slightly larger in outside diameter than the new pipe. This allows a small temporary annulus to be created between the replacement pipe and surrounding soil. The annulus created is very dependent on the type of soil surrounding the existing pipe. Depending on the original installation requirements of the existing pipe, compressible pipe backfill materials, such as sand or gravel, may also aid the bursting tool in expanding the area around the water main to create the annulus.

There is a select list of pipe bursting manufacturers that provide products designed for water main replacement. In addition, many of the manufacturers have several different models of pipe bursting tools and specialized accessories that accommodate the diameter and material of existing water main that is to be burst. There are basically two types of pipe bursting systems in currently use.

1. Pneumatic pipe bursting—compressed air–powered systems
2. Static pipe bursting—hydraulically powered systems

The choice of pipe bursting system is based on the material of the existing water main and the pipe material of the replacement water main being installed as well as soil conditions, depth of cover, length of runs, and proximity of other utilities.

## Pneumatic Pipe Bursting

In the pneumatic system, the bursting tool is a soil displacement hammer driven by compressed air. An expander is fitted to the rear flair of the pneumatic hammer. The pneumatic hammer assembly is launched into the host pipe via an insertion pit. The tool is connected to a constant-tension variable-speed winch cable that has been placed inside the existing water main being replaced located at the receiving point. The constant tension of the winch keeps the tool and expander in contact with the unbroken section of pipe and centered within the host pipe. The winch tension, when combined with the percussive power of the hammer, helps maintain the hammer and expander inside the existing pipe's center. The percussive action of the hammering and cone-shaped head is similar to hammering a nail into the wall; each hammer stroke pushes the nail a short distance. It cracks and breaks the existing pipe with each stroke. The expander combined with the percussive action push the fragments and the surrounding soil away providing space for the new pipe. See Figure 10-1 for a diagram of typical set up of pneumatic pipe bursting.

## Static Pipe Bursting

In static pipe bursting, pull force is applied to the expander through steel pulling rod assemblies inserted through the existing pipe. The expander transfers the horizontal pulling force into a radial force—breaking the existing pipe and expanding the cavity, providing space for the new pipe. With the rod method, steel rods are inserted into the existing pipe from the pulling shaft (receiving pit). The rods are connected together using different types of connections. When the rods reach the insertion pit, the bursting head is connected to the rods and the new pipe is connected to the rear of the expander.

A hydraulic power pack unit powers the static pipe bursting system pulling the rods one rod at a time. The rod sections are removed as pipe progresses. The expander and the new pipe are pulled in with the rod, fracturing the existing pipe and pushing the debris to the surrounding soil. The process continues until the bursting head reaches the receiving pit, where it is separated from the new pipe.

Properly designed and manufactured roller-blade cutting-wheel assemblies allow bursting of nonfracturing types of pipe such as steel and ductile-iron water pipes as well as repair clamps or fittings that may be on the pipeline. Due to the use of a bursting head or a roller-blade cutting-wheel assembly, static pipe bursting systems can burst both fracturable and nonfracturable host pipe materials. Static pipe bursting technology uses the pipe splitting method, which is essentially the addition of a *splitter* or *slitter* in front of the pipe burst expander head that splits the existing pipe (See Figures 10-2, 10-3, and 10-4 for diagrams of a typical set up of static pipe bursting).

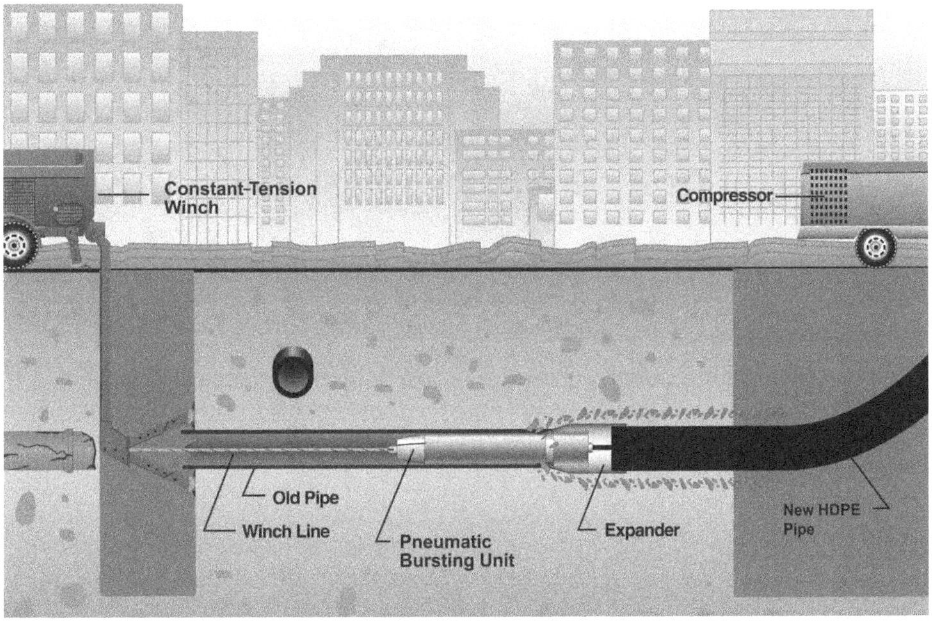


Courtesy of TT Technologies, Inc.

Figure 10-1 Typical pneumatic pipe bursting set up

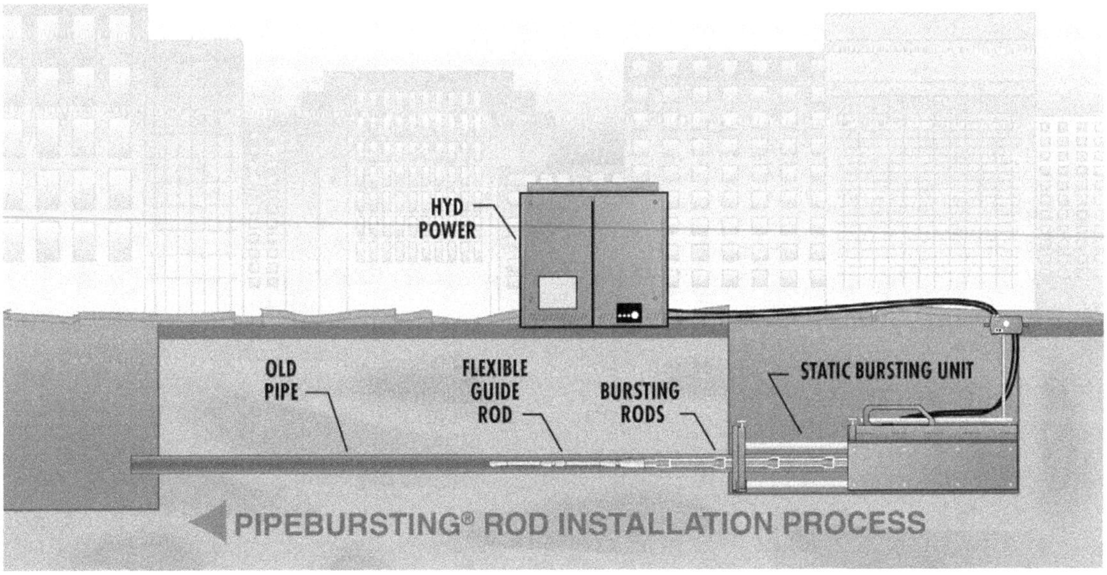


Courtesy of TT Technologies, Inc.

Figure 10-2 Typical static pipe bursting set up—step 1, bursting rod installation

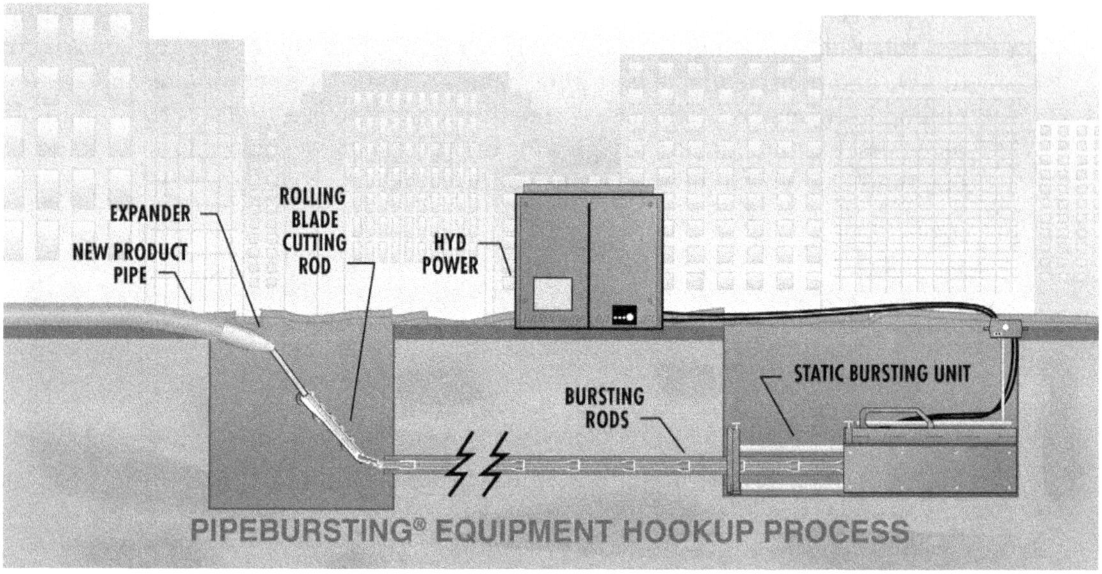


Courtesy of TT Technologies, Inc.

Figure 10-3 Typical static pipe bursting set up—step 2, bursting set up

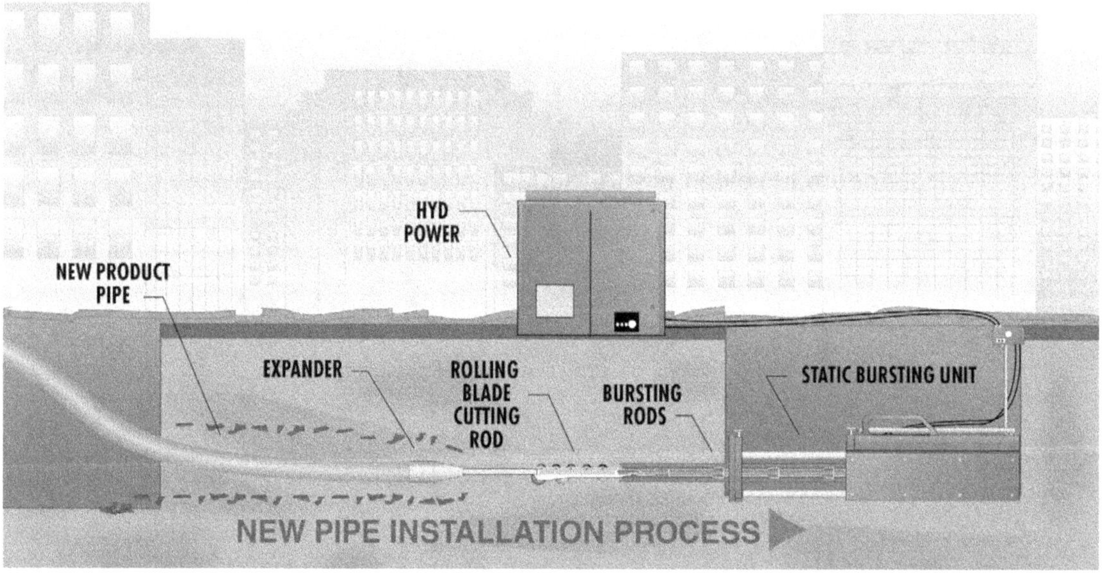


Courtesy of TT Technologies, Inc.

Figure 10-4 Typical static pipe bursting set up—step 3, pipe bursting

## WATER MAIN PIPELINES REPLACED BY PIPE BURSTING

Pipe bursting processes have been developed and improved to replace water main pipes constructed of

- Polyvinyl chloride (PVC)
- Ductile iron
- Cast iron
- Steel
- Reinforced concrete
- Asbestos cement
- Concrete

Virtually ALL existing water main pipeline materials can be replaced by the trenchless method of pipe bursting.

Most aging water mains have repair fittings or sections of pipe that have leaked and been replaced with short sections of pipe using mechanical couplings or 360° repair clamps. Special attention should be given to these repair sections. The static method of pipe bursting is the preferred method in water mains. The reason for this is because special mechanical attachments that will increase the chance to cut through unknown fittings can be attached to the static pipe bursting equipment that cannot be attached to the pneumatic systems. This will increase the success of splitting and allow for the expansion of these fittings. Special cutters should be specified that will improve the success in these cases (See Figures 10-5 through 10-8).

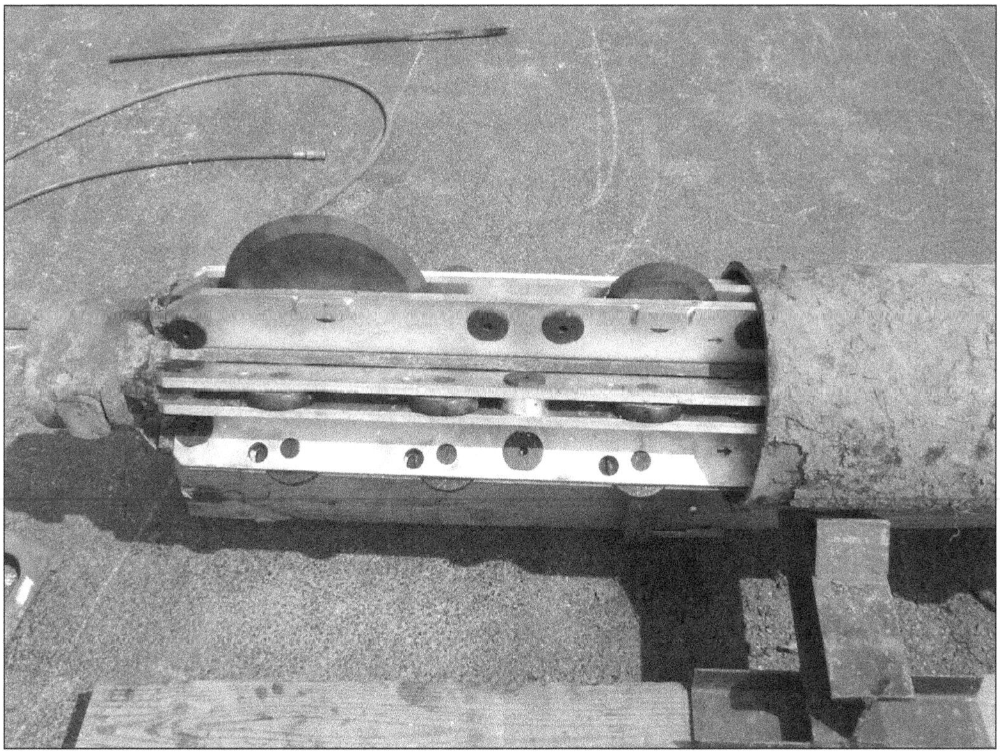

Courtesy of TT Technologies, Inc.

**Figure 10-5 Roller cutter used to cut through steel pipe and mechanical coupling**

Courtesy of TT Technologies, Inc.

**Figure 10-6 Roller cutter beginning to split steel pipe**

Courtesy of TT Technologies, Inc.

**Figure 10-7 Roller cutter splitting steel pipe while first cutter cuts barrel of mechanical coupling**

Courtesy of TT Technologies, Inc.

Figure 10-8 Roller cutter splitting not only steel pipe and barrel of mechanical coupling but also the ring sections of the fitting

## DIFFERENCES BETWEEN PIPE BURSTING

### Comparison with Traditional Open-Cut Replacement

Open-cut replacement may be a preferred option of pipe renewal when the pipeline is shallow and the trenching does not create inconvenience. However, under most conditions, pipe bursting has substantial advantages over open-cut replacements.

The advantages are especially notable in pipeline replacement for deeper lines, where the greater depth of lines increases the cost of open-cut replacement through additional excavation, shoring, and dewatering, etc., while the pipeline depth has minimal effect on the cost of pipe bursting. Water mains at depths of cover between four feet and six feet can also realize cost savings using pipe bursting over open cut depending on conditions and number of service laterals on the line. Additionally, as the underground utility network becomes even more congested through the advancement and expansion of gas, high-speed cable, and fiber-optic utilities, the need to preserve space underground for future growth becomes a necessity. By utilizing the existing utility corridor, new easements are not required and construction can take place through a previously opened trench. The grouting of the abandoned pipeline in place with a flowable-style grout is a standard practice when utilities are laid in a new location, which will only complicate future construction underground.

Additional advantages of pipe bursting over the open-cut replacement are indirect cost savings due to (1) less traffic disturbance, (2) shorter time for replacement, (3) less business interruption, (4) less environmental disturbance, (5) reduced surface paving expenses, (6) "green benefits" of lower carbon emissions, and other social benefits. Pipe bursting usually produces less ground disturbance than open replacement. In open cuts, there is stress relief in the ground as the trench is dug, and the unconfined ground moves inward and downward. Also, service lines parallel to the trench displace laterally and downward, while service lines crossing the trench sag. Shoring can reduce these movements but usually does not prevent them. Open-cut replacement that involves cutting through a road pavement structure can reduce the life of the pavement structure through backfill settlement in addition to the adjacent ground movements. Social costs such as traffic and business disruption, length of time and mess for open cut, reduced pavement life, environmental mitigation and others, all can increase the total effective cost of open-cut construction. In terms of contract price, when pipe bursting costs are the same or slightly more than a typical open-cut project, the decrease in total effective costs makes pipe bursting very attractive.

## Comparison with Other Rehabilitation Methods

Trenchless technology is a type of subsurface construction work that requires little or no surface excavation and no continuous trenches. It is a rapidly growing sector of the construction and civil engineering industry. Trenchless technology can be defined as, "A family of methods, materials, and equipment capable of being used for the installation of new or replacement or rehabilitation of existing underground infrastructure with minimal disruption to surface traffic, business, and other activities" (North American Society for Trenchless Technology 2012). Trenchless rehabilitation methods are generally more cost-effective than traditional dig-and-replace methods.

The pipe bursting method is proposed as a favorable alternative to sewer and water rehabilitation methods such as CIPP lining that relines an existing pipe with a liner that conforms to the profile of the existing pipe inside diameter (ID) while reducing it by the thickness of the lining material installed. While relining methods offer no-dig rehabilitation of an existing pipe and they follow the grade and profile of the existing pipe, pipe bursting can install a new pipe with a true profile (design ID). This ability can be advantageous, if used to correct offset joints or deflections in the existing pipe. Areas requiring corrections are places where pipe bursting launch and receiving pits can be placed and taken advantage of.

One significant advantage of pipe bursting over other trenchless rehabilitation methods, such as CIPP lining, FIPP lining, sliplining, spirally wound pipes, etc., is the ability to upsize existing underground pipelines, thus increasing hydraulic capacity by use of a trenchless method. Pipe bursting is the only trenchless method that can increase the hydraulic capacity of a pipe by installing a new pipe of the same or larger ID. In mainline sewer and water main pipe bursting, additional reduction of inflow and infiltration from the sewer service connections and reduced water loss from water service connections are added benefits as they are reinstated manually with a new hard connection to the mainline; whereas CIPP or other methods will only reinstate the existing service connection from the main robotically or by other means from inside the mainline. Long-term benefits of reconstruction of the main and service connection can be significant when comparing trenchless rehabilitation methods to a trenchless replacement method like pipe bursting.

Cement–mortar lining and polymeric linings are methods that allow water utility systems owners to clean and reline existing cast-iron, steel, or ductile-iron pipe with a centrifugally cast cement–mortar or epoxy lining, which restores the original pipe hydraulic capacity and reduces further corrosion; however, these methods offer little structural support for the aging pipe. Extensive cleaning of the pipe to remove encrustation is required whereas with a static pipe burst system, the solid rod design allows the insertion through heavily encrusted pipes with little or no cleaning, thus reducing the preparation cost and downtime of the system. Pipe bursting will allow a new manufactured pipe to be installed that can have a larger diameter, thus dramatically increasing the pipe flow characteristics and potential operating pressure. The newly installed water main can then undergo pressure tests up to 150 psi dependent on new pipe type and pressure class installed, thus providing a long-term structural pipe replacement.

Prechlorinated pipe bursting is an alternative to setting up temporary water service lines and allows for the newly prepared pipe to be pressure tested, disinfected, and accepted before installation allowing rapid installation and reconnection, often in one day. Pipe bursting may be the only choice for trenchless improvement of an existing pipe in very poor structural condition or if other rehabilitation methods are rejected as unsuitable.

## PROJECT EXECUTION RECOMMENDATIONS

As mentioned in chapter 2, Preconstruction Activities, with any pipeline construction project, a thoroughly planned and detailed design will frequently lead to a successful construction barring any unforeseen issues. A project design of plans should identify existing and any proposed service connections, existing valves, hydrants, and fittings, so they can be located, excavated, and exposed before pipe bursting operations commence. A geotechnical report that evaluates the existing soil characteristics (type, compaction, strength, etc.) along the water main alignment should be prepared. The report should provide information on the soil conditions above the existing water main and outside of the trench so that information can be included in the project specifications.

Once the information has been collected on the existing conditions, the designer should contact several pipe bursting tool manufacturers to discuss the project, its conditions, and obtain recommendations on which tool to specify for the work or develop a list of concerns so that the client agency is aware of any potential issues that may arise during construction. This will foster a better educated project environment.

During construction of the water main replacement, a temporary potable water bypass system most likely will be required to maintain water service to existing customers. This process is routine for the agencies and contractors that are experienced in this work. Many think that installing a temporary bypass system, and excavating at each service connection would make pipe bursting costs closer to open cut. On the contrary, even with these additional services, costs for pipe bursting have saved on average between 15 percent–45 percent over open-cut construction depending on the conditions. The details regarding the bypass or project/community communication responsibilities should be outlined in the project contract documents so the contractor is aware of any responsibilities it is required to perform. Refer to chapter 3 for additional details on bypass systems and chapter 13 for communicating the project with the impacted customers.

Once all of the water service connections are located, they should be excavated, completely disconnected, and isolated from the existing water main before pipe bursting operations commence. Service connections should not be reconnected to the replacement water main pipeline until installation, pipeline relaxation (if required by pipe manufacturer), disinfection, and pressure testing are complete.

All other utility pipelines and underground structures within close proximity of the water main to be replaced using pipe bursting should be excavated (or *pot holed*) and exposed prior to performing the work. Depending on the existing soil characteristics and the resulting diameter of the pipeline to be installed, additional distance may be necessary to prevent damage to a nearby utility or structure from the installation process. It is highly recommended that the soil between the existing pipe to be burst and other structures be removed so bursting forces are not transmitted through the soil to the adjacent structures, reducing any potential for unintentional damage.

Cleaning of existing pipe interior is not necessary for most pipe bursting processes unless it is not possible to pass the pulling cable or push rods through the existing pipe. In cases of severe encrustation or internal obstacles, some cleaning may be required. For typical water pipelines, any existing biofilm or sediments in the existing pipe are simply pushed into the surrounding soils.

Access excavations are needed to install the replacement water main sections. Access excavations are required along the existing pipeline alignment to provide for removal of all existing valves, fittings, and other known obstructions from the project site. Access excavation paving, excavated soil, and vegetation should be removed and legally disposed of as with any water main replacement project. Access excavations will also be required at the beginning and end of the water main to be replaced unless other locations are required. Depending on the configuration of the existing water main (abrupt changes in direction, length of the proposed pipe burst, etc.), intermediate access excavations may be required. Excavations should be centered over the existing water main, and excavation requirements should be verified in the field and confirmed with the client prior to construction of the project.

## REPLACEMENT PIPE MATERIALS

Currently, there is a wide array of replacement water main pipe material choices available for pipe bursting installations: high density polyethylene (AWWA Standard C 906); restrained joint ductile-iron pipe (AWWA Standard C 151); Fusible FPVC (AWWA Standards C 900 and C 905); and restrained joint PVC (AWWA Standard C 900) are commonly used.

Historically, polyethylene material is a product that was used for replacement of water mains. Sections should be assembled and joined on the job site using the butt-fusion method. The butt-fusion process is performed in the field by qualified operators with prior experience in using the proper jigs and tools according to procedures outlined by the pipe manufacturer and the butt-fusion equipment manufacturer. Every butt-fused joint should have a smooth, uniform, double-rolled back bead made while applying the proper melt, pressure, and alignment during the fusion process. Care must be taken not to damage the replacement water main as it passes through the burst fragments of the original pipe. The replacement water main pipe should be made available for inspection by utility staff before the pipe bursting operation begins. The same butt-fusion process is used to weld FPVC pipe.

Other pipe materials, such as jointed ductile iron or restrained joint PVC allow the replacement water main sections to be assembled one length at a time during the pipe bursting operation. This process is known as *cartridge loading*. One benefit is that this process will allow the replacement water main to be installed without lengthy lay-down areas that are required for butt-fusion welded types of pipe. This is a good application in very urban settings or sites that do not allow for long pipe lay-down areas.

Similar to an open-cut construction water main replacement project, the contract documents should clearly outline the acceptable or agency-approved water main replacement materials requirements and whether substitutions will be allowed.

Pressure testing and disinfection should follow installation of the replacement pipe according to the same procedures used for any water main installation (refer to ANSI/AWWA Standard C651 for a detailed description of the requirements). After the replacement pipe has been completely installed, disinfected, and pressure tested, all existing active services indicated on the contract plans or identified by the utility should be reconnected to the replacement pipe.

Disinfection options:

1. Chlorination in accordance with ANSI/AWWA Standard C651 and/or local agency requirements, similar to open-cut construction.
2. Prechlorination. This process applies to replacement materials that are fused and allows for the new replacement water main to be joined together and then chlorinated, tested, and certified, then put into service immediately after installation. The string of prechlorinated pipe is installed in a normal pipe bursting operation and then the pipe is put into service the same day.
3. One important aspect of the construction of replacement water mains is to require a licensed and experienced contractor. The pipe bursting process is a specialized construction process and therefore the contractor who is responsible for the pipe bursting services should provide proof that they have been trained by the manufacturer in the proper use, operations, and maintenance of the pipe bursting equipment. In addition, it is important to retain the services of a contractor who has experience in the installation of a water main using pipe bursting technology with similar project characteristics (soil type, diameter of replacement pipe, existing and replacement water main pipe material, etc.)

A project requirement that should be included in the project contract documents is that the contractor submits written descriptions of the construction methods and manufacturer of the pipe bursting equipment to be used.

## CONCLUSIONS

Pipe bursting is a mature technology with a proven history for trenchless replacement of existing water mains. As the life cycle of the existing underground infrastructure expires and failures occur at an alarming rate, pipe bursting is one of the methods that will be used effectively to provide long-term service for critical utilities that are essential to public life and health. As the only trenchless method that can increase the size of the existing pipe, while offering a variety of factory manufactured pipe materials to be installed, pipe bursting is suited well to a growing need for additional capacity whether it be in the sewer, water, gas, or other utility market sectors. With increased public awareness and limited funding available for critical infrastructure rehabilitation, utilities should use methods that reduce social disruption and environmental impact, and prepare for future capacity needs by leveraging technology that provides a better product both the short and long terms.

As with any successful construction project, pipe bursting projects require good preplanning, careful observation of job progress, manufacturer and contractor experience, and key monitored variables during construction. The result will be a good installation providing additional capacity and services to the owner and community for a multitude of years.

## REFERENCES

American Water Works Association (AWWA). 2005. Standard for Disinfecting Water Mains. ANSI/AWWA C651. Denver, Colo.: AWWA.

AWWA. 2007. Standard for Polyvinyl Chloride (PVC) Pressure Pipe and Fabricated Fittings, 4 In. Through 12 In. (100 mm Through 300 mm), for Water Transmission and Distribution. ANSI/AWWA C900. Denver, Colo.: AWWA.

AWWA. 2010. Standard for Polyvinyl Chloride (PVC) Pressure Pipe and Fabricated Fittings, 14 In. Through 48 In. (350 mm Through 1,200 mm). ANSI/AWWA C905. Denver, Colo.: AWWA.

AWWA. 2012. Standard for Polyethylene (PE) Pressure Pipe and Fittings, 4 In. Through 65 In. (100 mm Through 1,600 mm). ANSI/AWWA C906. Denver, Colo.: AWWA.

Freed, T., B. Metcalf, G. Mallakis, A. Mayer, E. Nicholson, B. Botteicher, M. Rocco, M. Timberlake, M. Werth, and M. Woodcock. 2012. *International Pipe Bursting Association Guidelines for Pipe Bursting.* Marriotsville, Md.: National Association of Sewer Service Companies and International Pipe Bursting Association.

North American Society for Trenchless Technology. 2012. *Trenchless Technology Definition.* http://nastt.org.

Chapter 11

# Reinstatement of Service Laterals

How service laterals are reinstated can have a significant influence on project costs and schedule. The effort can vary greatly, depending on the type of rehabilitation that is employed, and the method that is used for reinstatement. For the spray-applied linings, the reinstatement of service laterals is very simple. For other types of rehabilitation, service lateral reinstatement may involve either external excavations or internal pipeline "robots" that are remotely controlled. Significant technological advancements in this area have occurred over the last few years.

## LATERAL REINSTATEMENT FOR SPRAY-APPLIED LININGS

### Cement–Mortar Lining

For cement–mortar lining, service reinstatement is very simple. In this time-proven method, the only concern is re-establishment of the lateral opening, i.e., making a hole in the lining at the lateral connection. Commonly, the cement mortar will plug the lateral opening during its application and would obstruct the flow of water to the customer. This blockage is routinely prevented by a blast of compressed air or water that is directed down the lateral from the meter end of the pipe, before the mortar has fully set. Because the lining is primarily intended to protect the inner surface of the pipe from corrosion, a positive connection between the lateral pipe and the lining is not necessary. Any water that enters the annulus between the mortar and the pipe will become highly alkaline and noncorrosive.

Any new services that are tapped into a water main rehabilitated with cement–mortar lining can be accomplished in the same manner as any main constructed of factory lined pipe. The tools and techniques needed are familiar to nearly all water utility crews.

### Thin Polymer Linings

Polymer linings (epoxy, polyurethane, and polyurea) are most commonly applied at a thickness of about 0.039 in. (1 mm). At this thickness, service reinstatement is even simpler; the lateral opening is not generally plugged by the lining, so no effort is needed to re-establish the opening. Like cement–mortar lining, these thin linings are intended primarily to protect the inner surface of the pipe from corrosion, and no direct connection between the lateral pipe and the lining is required, provided that the lining is well adhered to the pipe.

New services that are tapped into a water main rehabilitated with a thin polymer lining can likewise be accomplished using standard tools and techniques. In some cases, a hole saw or very sharp drill bit is recommended, to assure the lining is cut cleanly and not detached from the pipe wall.

### Thicker Polymer Linings

The development of fast-setting polymers (polyurethanes and polyureas) has allowed the application of pipe linings that are thicker than the standard 0.039-in. (1-mm) lining. Recently, linings ranging from 0.12 to 0.197 in. (3 to 5 mm) in thickness have been applied, with the intention that these thicker linings have greater ability to span weaknesses in the host pipe, as a Class II lining. However, in their application, these thicker polymer linings can plug lateral openings, particularly if the opening is on the side of the pipe. As with cement–mortar lining, a routine blast of compressed air can be directed down the lateral to assure that such plugging does not occur.

When tapping new services into a water main that is lined with a thick polymer lining, the need for appropriate hole saws or drill bits is important, to prevent detachment of the lining from the pipe wall.

## LATERAL REINSTATEMENT FOR NONSPRAY-APPLIED LININGS

For linings that are not spray-applied, reinstating service laterals generally requires the need for either an exterior excavation or an internal pipeline robot. For these semi-structural and structural linings, three basic steps are required in lateral reinstatement: (1) finding the lateral; (2) re-establishing the lateral opening; and (3) connecting the lateral opening to the lining or carrier pipe.

1. **Finding the lateral.** For exterior connections, finding the lateral is generally rather simple; standard pipe locating devices are used. For interior connections, however, the issue is more complex. Nonsprayed linings cover-up the lateral openings, often making them difficult to locate from within the pipe. A closed-circuit television camera can sometimes be used to find a bulge in the lining if the service tap protrudes into the pipe, but unlike with wastewater laterals, a dimple in the lining is generally not seen at a service connection because the diameters are too small. To overcome this, sophisticated techniques have been developed, which map the laterals before the lining occurs or detects them afterward. One company uses an electromagnetic nondestructive technique to detect the lateral locations beneath the lining, as described later in this chapter.
2. **Re-establishing the lateral opening.** For exterior connections, this step is usually simple and generally involves standard tapping tools and techniques. However, for an interior connection, this step is more complex. Although several different pipeline robots have been developed to cut holes into wastewater liners in water

mains, extreme precision is needed; the hole in the lining must align perfectly with the lateral bore, which is generally quite small. As a consequence, special robots are needed for this work.

3. **Connecting the lateral to the lining or the carrier pipe.** For most nonsprayed lining systems, simply re-establishing the lateral opening is not sufficient. If the lining is intended to be structural or semistructural (i.e., Class II, III, or IV), leakage of water into the space between the lining and host pipe must be prevented. If leakage into this annulus occurs, the structural benefits of the lining are negated, as well as the corrosion protection benefits. Therefore positive connections between the lining or new carrier pipe and the lateral pipe are required. Such positive connections can be accomplished in three basic ways:

    a. **External reconnection.** This is the traditional method. A pit is excavated at each lateral in order to make a new mechanical connection between the lateral and the lined pipe or new carrier pipe. A traditional pit must be large enough for a person to work in, roughly 3 ft by 5 ft in plan; however, *keyhole* techniques, as described later in this chapter, can also be used, which require much smaller excavations. How the connection is made depends on the material used for the lining/carrier pipe. If HDPE is used, for example, an electrofused, self-tapping saddle is generally used.

    b. **Lining adhesion to host pipe.** If a cured-in-place lining is well adhered to the host pipe at the lateral connection, a positive connection may already exist between the lining and the lateral. With good adhesion, the lining is bonded to the host pipe, which in turn is connected to the lateral (through the existing tap). Good adhesion requires a clean, dry substrate and an appropriate adhesive material.

    c. **Internal tap.** Where a modified (or close-fitting) sliplining method is used, a mechanical connection between the lining and the lateral can also be made from inside the pipe. This seal is accomplished using a self-tapping insert that screws into the existing tap. The insert is equipped with a compressible washer that seals the insert to the lining. The device is installed using a remotely controlled pipeline robot. Although developed for close-fit sliplining methods, this method could also be used for cured-in-place lining. Figure 11-1 illustrates an internal tap.

In the rehabilitation of gravity wastewater mains, the sealing of lateral connections is also accomplished using various grouts, sealants, and cured-in-place inserts installed from within the pipe. These have not yet been successfully adapted to water main applications, where pressures are typically much higher. A special concern for water systems is avoiding harmful leachates in the water. (All such grouting and sealing materials should be tested and certified in accordance ANSI/NSF Standard 61.)

## Keyhole Methods for Lateral Reinstatement

When pipe bursting or loose-fit sliplining methods are used, external excavations will be required to connect the lateral pipes to the new carrier pipe. To reduce the impact of these excavations on the community, to possibly reduce the project cost, and to shorten the project schedule, keyhole construction methods should be considered.

Keyhole construction entails accomplishing work from the surface of the street, using long-handled tools inside a small excavation. Keyhole construction takes the place of traditional construction involving workers inside larger excavations. By using keyhole methods, the amount of excavation and backfill is reduced, the need for shoring is eliminated, small traffic plates can be used (installed and removed without a backhoe), and pavement repair costs are much reduced. A typical footprint for a lateral connection made with keyhole techniques is 18 by 24 in. in diameter. Figure 11-2 shows various keyhole tools that are used to accomplish the work.

A keyhole lateral reconnection is accomplished as follows:

1. The services lines and mains are traced using pipe locating equipment, and the locations of the main/lateral connections are marked on the pavement. The limits of the demolition are then marked.
2. The pavement is broken or cut using saws, jack hammers, or coring tools.
3. Vacuum excavation equipment is used to excavate the keyhole, centered over the corporation stop. The top and sides of the main are exposed.
4. After the bypass system is in service and the main is dewatered, the existing connection between the main and the lateral is severed. (If the pipe bursting method is being used, this step needs to occur prior to insertion of the pipe bursting tool, to prevent damage to the lateral.)
5. A small traffic plate is placed over the excavation, pending completion of pipe bursting or sliplining operation.
6. After the new carrier pipe is installed, long-handled tools are used to complete the connection of the lateral pipe to the new main. Figure 11-3 shows such a connection prior to its installation.
7. After pressure testing is completed, the excavation is backfilled and the pavement is repaired. Due to the small size of the excavation, the use of slurry backfill is appropriate.

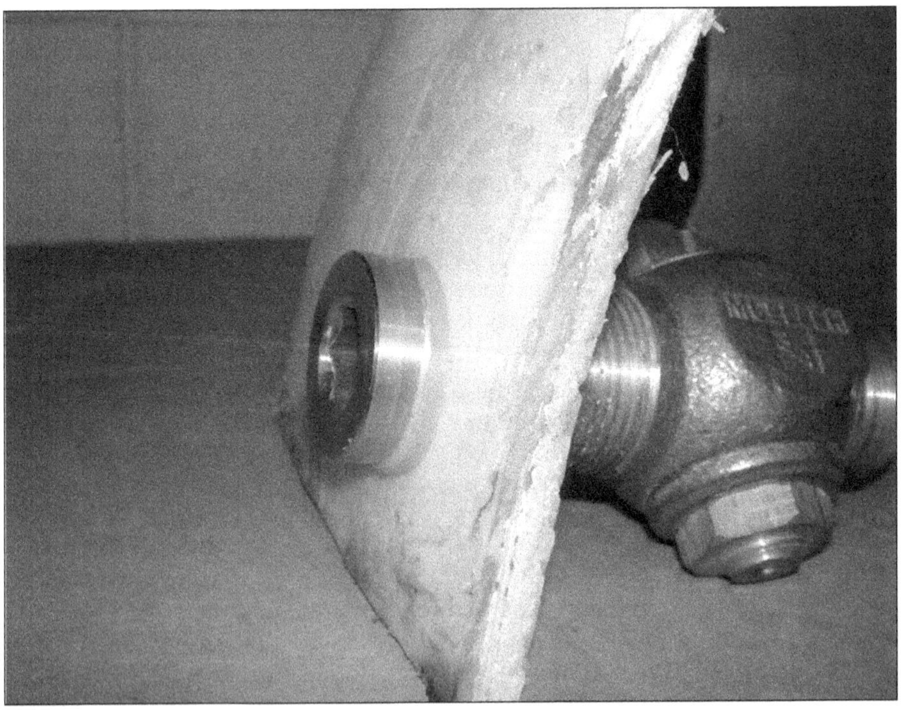

Courtesy of Insituform Technologies, LLC

**Figure 11-1 Internal tap connecting a structural or semistructural lining to a lateral pipe**

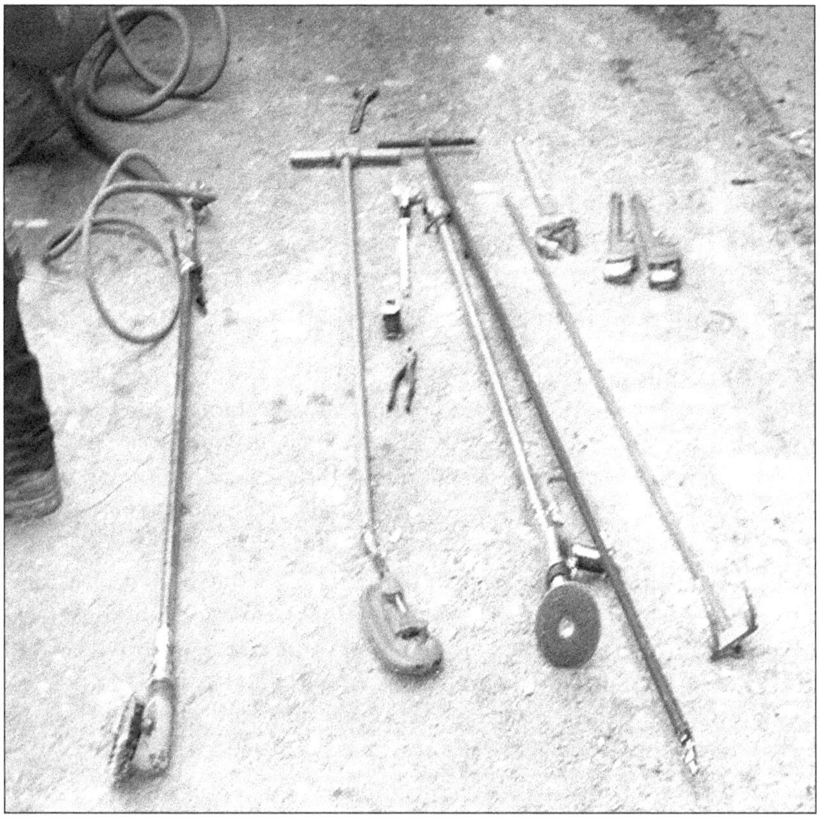

Courtesy of Water Research Foundation

Figure 11-2 Various keyhole tools used to connect service laterals to a new water main

Courtesy of Water Research Foundation

Figure 11-3 Electrofusion saddle, with copper pipe and fittings, ready for installation with long-handled tools

## PAVEMENT CORING AND GROUTING

When employing keyhole techniques, pavement coring should also be considered. By coring rather than cutting the pavement, stress concentrations are avoided and less damage to the pavement occurs (Figure 11-4). Then by retaining and reinserting the extracted coupon, the road surface can be quickly repaired with fast-setting grout, with little permanent evidence that the pavement was cut. The pavement repair can be ready for traffic within 30 min of grouting.

## PIPELINE ROBOTS FOR LATERAL REINSTATEMENT

When a tight-fitting structural or semistructural lining is used for rehabilitation, pipeline robots can be used to find the service laterals, re-establish the lateral openings, and install connections from the liner to the laterals. The method described below was developed by one rehabilitation company for reinstating services following close-fit HDPE lining. Another robotic method has been developed by another company for service reinstatement after reinforced cured-in-place pipe lining.

1. After the bypass water supply system has been placed in service, but before the water main is lined, a pipeline robot prepares the existing corporation stops by milling off any protrusions into the main, and by boring the interior of the corporation stop to a uniform inside diameter. Boring and milling must be done in a counter-clockwise direction to avoid loosening of the corporation stop. The location of each corporation stop is also recorded.
2. After the lining has been installed, remote field electro-magnetic technology (also called *eddy current* technology), is used to precisely detect the location of each corporation stop.
3. Using a milling tool, the pipeline robot cuts away the lining covering the corporation stop.
4. A self-tapping insert is then screwed into the corporation stop by the pipeline robot. The insert is essentially a short nipple piece with a flange on one end and exterior threads on the other. The threads are reverse-threaded, so that tapping into the corporation stop does not unscrew the corporation stop from the main. Under the flanged part of the insert is a compressible doughnut-shaped rubber seal. As the insert threads into the corporation stop, the rubber seal flattens against the inside wall of the lining, forming a water-tight seal.

Courtesy of Utilicor Technologies, Inc.

**Figure 11-4 Keyhole core and replacement coupon**

## REFERENCE

American National Standards Institute (ANSI). 2012. Drinking Water System Components—Health Effects. NSF/ANSI 61-2012, 198. Ann Arbor, Mich.: NSF International.

This page intentionally blank.

Chapter 12

# Cathodic Protection Retrofits

Corrosion is an electrochemical process in which metals are reduced to their more natural oxidized state. Corrosion control requires the ability to change and monitor changes in the electrochemical state of the pipe surface. Proper design and implementation of corrosion control to a structure allows cost-effective control and monitoring with a minimum amount of disruption to the operation of the pipeline and interference to adjacent structures.

The water utility industry is inherently different as compared to the oil and gas industry. The water industry is not regulated to apply cathodic protection to its pipelines. Application of cathodic protection in the water industry is elective and only for the purpose of leak rate reduction and asset preservation.

Corrosion-control recommendations are different between new and existing pipe. For existing systems where selection of materials and application of coatings is not possible, cathodic protection is an effective means for reducing corrosion-related problems. The application of cathodic protection implies a commitment to maintaining a corrosion monitoring program to ensure that the cathodic protection systems continue to provide the necessary level of protection and that protection does not diminish over time.

## PREDESIGN FIELD TESTING

### Pipe-to-Soil Potentials

Pipe-to-soil potential measurements are the cornerstone of the corrosion engineer's arsenal. Potential measurements can be used to assess the likelihood of active corrosion on unprotected structures, stray current from foreign pipelines, and to assess the effectiveness of cathodic protection systems. Critical evaluation of the potential data is necessary in order to extend the life of a pipeline and add or adjust corrosion mitigation systems as warranted.

Proper evaluation of a cathodically protected structure is just as important as evaluating noncathodically protected pipe-to-soil potentials. NACE International Standard SP0169-07 provides three criteria for evaluating the effectiveness of a cathodic protection (CP) system using pipe-to-soil potentials: (1) a negative potential of at least 850 mV relative to a saturated copper/copper sulfate reference electrode with CP applied; (2) a negative polarized potential of at least 850 mV relative to a saturated copper/copper sulfate reference electrode; or (3) a minimum of 100 mV of cathodic polarization between the structure's surface and a reference electrode contacting the electrolyte. The selection of the proper pipe-to-soil potential criterion for a cathodically protected structure is an issue that should be considered by an owner. For example, by changing criteria, rectifier current capacity, anode life, and economic benefits may be extended and lessen the potential stray current impact on adjacent utilities.

## Soil Testing

A major factor in determining soil corrosivity is electrical resistivity. The electrical resistivity of a soil is a measure of its resistance to the flow of electrical current. Corrosion of buried metal is an electrochemical process in which the amount of metal loss due to corrosion is directly proportional to the flow of electrical current (DC) from the metal into the soil. Corrosion currents, following Ohm's Law, are inversely proportional to soil resistivity. Lower electrical resistivities result from higher moisture and soluble salt contents and indicate corrosive soil. Several techniques are used for corrosivity testing, which should be performed by trained and certified professionals. Figure 12-1 shows one technique used for over-the-line corrosivity testing.

Courtesy of HDR Engineering, Inc.

Figure 12-1 Over-the-line electrical resistance testing

### Electrical Continuity Testing

Pipeline electrical continuity is necessary to facilitate not only a functional cathodic protection system, but also to mitigate the causes of stray current interference, and provide adequate levels of monitoring and testing. Electrical continuity of a pipeline may be determined by utilizing one or more of the following procedures: (1) circulating current through segments of a pipeline and comparing pipe conductance results against theoretical values; (2) application of a temporary cathodic protection system and monitoring the span's response; (3) application of temporary cathodic protection current and measuring earth gradients between two electrodes; (4) interrupting existing cathodic protection systems and monitoring the current flow at test stations; (5) tracing the pipe alignment with a suitable, tuned pipeline locator; and (6) measuring current with a pipeline current mapper.

### Cathodic Protection Current Requirement Testing

The most accurate and desirable method for determining the amount of cathodic protection current required for protecting a structure is to measure the actual amount of current required to achieve protection through installation and operation of a temporary cathodic protection system. Measurements of the currents supplied and the structure-to-electrolyte potentials responses are used to calculate the required current to protect the structure.

## SYSTEM DESIGN

Cathodic protection current for a pipeline may be provided by galvanic anodes or impressed current systems. Generally, galvanic anodes are used where the current requirement is low and soil resistivities are less than 3,000 ohm-cm. Galvanic anode material for buried or submerged applications include aluminum, magnesium, and zinc. Soil or water resistivity and chemistry, and the application will dictate which type of anode material is required. Often, galvanic anodes are applied where pipe repairs are performed. Even where electrical continuity is limited, the anodes can protect a portion of pipe where problems have occurred.

Where soil resistivity is high or current requirement is large, impressed current systems are designed. Impressed current anode materials include graphite, high-silicon cast iron, mixed metal oxide, platinum, and titanium.

The current requirement for impressed current systems is measured by field testing for existing pipelines or estimated with applicable and approved current requirement densities for new pipelines. Rectifiers should be sized such that the amperage rating provides the required current with an allowance for future coating deterioration. Figure 12-2 shows a corrosion engineer field-verifying voltage and current requirements.

Impressed current anodes are rated by the manufacturer for life expectancy at a known current output. The size and number of anodes are selected based on measured or estimated current requirement and manufacturer data.

Other design factors for cathodic protection to be considered include, and are not limited to, insulated joints; test stations; permanent references electrodes; joint bonding; remote monitoring; backfill; right-of-way restrictions; local, state, or provincial regulations; availability of AC power; owner preference; interference, power costs; and ease of replacing consumed anodes.

Courtesy of HDR Engineering, Inc.

Figure 12-2 Verifying voltage and current requirements for a cathodically protected pipeline

Project drawings and specifications should be reviewed by a qualified corrosion engineer to determine if corrosion control recommendations made as a result of preliminary document review, corrosivity evaluation, and field testing are properly incorporated. Any changes based on the review should be reviewed again by qualified personnel in order to maintain project QA/QC requirements.

## TESTING AND MAINTENANCE

Testing and inspection should be performed during construction and post-construction, but before the construction contract is complete in order to determine if the contractor has properly installed the designed corrosion-control features.

Cathodic protection systems in general are not maintenance intensive by design. Galvanic anodes require regular annual inspection by operations personnel. Impressed current systems typically require a monthly log of rectifier outputs (voltage and current). This simple monitoring activity for impressed current systems minimizes the possibility that the system is inadvertently turned off or becomes inoperative. New impressed current systems can be installed with integrated remote monitoring and control systems that allow data collection and adjust via modem or satellite uplink from a central location.

A periodic survey of both galvanic and impressed current systems by qualified personnel or contractor is recommended to verify and adjust, if necessary, the level of protection. The frequency for surveys should be based on a review of historical information and data and owner's preference and budgetary constraints.

Data collection is the first step to developing a database of corrosion-control information that forms the basis for developing a long-term corrosion maintenance and control program.

## REFERENCE

National Association of Corrosion Engineers (NACE). 2007. Control of External Corrosion on Underground or Submerged Metallic Piping Systems. NACE SP0169-2007:35 Houston, Texas: NACE.

Chapter 13

# Program Management

Program management of any water supply and distribution rehabilitation project includes identification of the customer relations issues, making provision in the contract documents to provide good customer service and to measure the benefits gained after the project is completed. Certain measures are necessary during the planning and execution of the project that must be understood very clearly by the utility before embarking on such projects.

## CUSTOMER/COMMUNITY RELATIONS

Perhaps one of the most important long-term operational or maintenance aspects of any water supply system is the relationship that develops between the utility and its customers, in part through the utility's efforts to maintain continuous and reliable water service. In this regard, both residential and commercial/industrial customers must understand the general nature of any project that may affect their water supply. This information should be provided to the customers in a timely and effective manner prior to a scheduled project. Advance notice allows customers to ask questions and resolve problems prior to beginning the actual rehabilitation work. Requiring the rehabilitation contractor to participate in the notification process is also beneficial. The name and contact information of the rehabilitation contractor should be provided during the information exchange. Planning and implementing a good customer relations program is equally as important as planning the actual rehabilitation work.

Depending on the water agency preference, if a water system bypass is not utilized during the rehabilitation lining technique, a boil-water advisory may need to be employed depending on the circumstances.

## PROJECT NOTIFICATIONS

Notifications irrespective of the type should stress the necessity of the project and explain what will happen during construction. A letter describing the project, including details of detours and local area access, is recommended (Figure 13-1). Letters should be delivered to all residential and commercial customers within and near the project area approximately a week in advance of construction. Notifications to the customers requiring special attention must be delivered separately and acknowledgments secured. Additionally, doorknob hanger cards that explain the project and that provide outage notification and dates can be used immediately before the impacting work is set to begin. Customers should be notified about outages a minimum of 48 hr for businesses and 24 to 36 hr for residents (Figure 13-2). In addition, a utility should establish an appropriate section on their web page that is updated frequently with project and construction information for customers to keep informed.

Emergency service providers such as paramedics or ambulance companies should be notified of the project. Other public and investor-owned utilities, such as sewage, gas, electric service, and telecommunications (telephone, cable, and fiber-optics cable) should also be notified to allow coordination between the water line project and any other scheduled utility work. In all notifications, a detailed map of the work and a proposed construction schedule should be provided to the agencies. Any changes in the project schedule or location are just as important to the public and affected agencies as the original information and should be sent out as soon as the change is known. The local chamber of commerce or other local business groups may be contacted for assistance in reaching out to the business community about the project, its schedule, and the impacts and benefits. Notices posted in store windows may help to avoid parking complaints for the businesses if the project is in a commercial area. Local government representatives should also be made aware of the project and its details.

All notifications should include a live emergency telephone number that is active 24 hours a day, 7 days a week. The project specifications should require the contractor to provide an emergency response person and phone number. Good customer relations can be severely damaged by a customer's inability to report an emergency or discuss questions about water service. Depending on the size of the project and the number of impacted customers, a contact person for the general public other than the construction superintendent may be necessary. If a project requires entering private property to disconnect meters and hook up temporary service lines, notices should clearly discuss the need for the work, the schedule, and how the work will be performed. This coordination will prevent unforeseen circumstances, misunderstandings about responsibilities, and any possible unfriendly family pets.

As with any aspect of customer service, follow-up is extremely important and personnel must be available to respond promptly to all calls even during off hours. Finally, an established policy and procedure for handling damage claims should be established prior to beginning of the rehabilitation work. A survey of the area and/or video documentation of private property conditions prior to the commencement of work on private property are highly recommended. A customer satisfaction survey at the conclusion of the work will reveal any problem areas and possible improvements for future work.

**Hackensack Water Company**

200 Old Hook Road
Harrington Park, N.J. 07640

PURVEYORS OF PURE WATER • SINCE 1869 •

**Dear Customer:**

Starting the week of July 16, Hackensack Water Company will be cleaning and lining a major water main which supplies portions of your community. Following the route indicated on the map below, construction will begin in Englewood at the intersection of Van Nostrand Avenue and Grand Avenue, and will proceed south on Grand Avenue through Leonia, Palisades Park, and Ridgefield, to end at the Hendricks Causeway. The project is expected to take three to four months to complete.

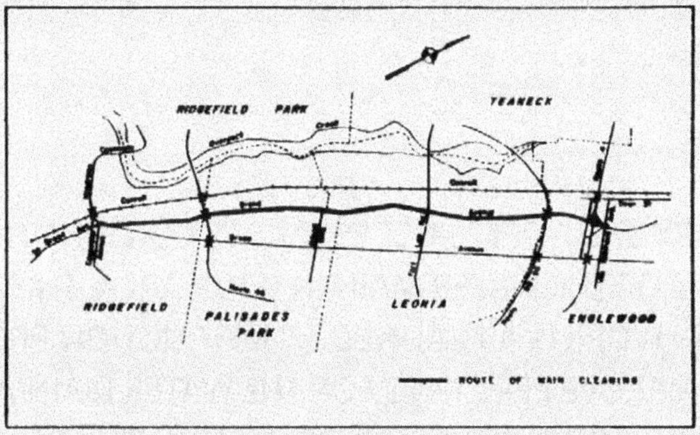


## IMPROVING YOUR WATER SYSTEM

As part of our ongoing maintenance program, we must periodically recondition sections of our water system, which assures you, as our customer, continued dependable water service. Pipeline reconditioning involves scraping the inside of the main to remove accumulated materials, followed by the application of a thin coat of cement to protect the water main from further corrosion. This work results in improved water flows in the pipeline and more efficient operation of our system. As a customer, your benefit is the assurance that water pressure in your home or business, as well as essential fire protection, will be maintained at all times.

## HOW WILL THIS WORK AFFECT YOU?

While the work is in progress, residents and businesses along the route will be provided with a temporary water supply since the water main must be taken out of service for cleaning. Connections will be made to your service line from alternate pipelines laid alongside the curbs on each side to the street. We will make every effort to ensure that your normal water service is maintained at all times.

## PLEASE...WATCH YOUR STEP

If you live, travel or shop in the area, you'll notice a number of openings in the street to allow access to the water main and to accommodate the bypass pipelines running across the intersection and driveways. All openings will be covered during the work and will be resurfaced as soon as possible after the project is finished. Every possible safety precaution will be taken and all work areas will be barricaded and carefully marked to prevent injury. You can help by cautioning your family and friends to watch their step and avoid the work areas as much as possible.

## INCONVENIENCE TEMPORARY...IMPROVEMENTS PERMANENT

We realize our main cleaning project may cause some inconveniences for those who live and work in the area. We are working closely with your local officials and police department to minimize traffic disruptions and we will do our best to see that the construction goes smoothly.

## IF YOU HAVE A PROBLEM...

If a problem should occur as a result of our work, please report it directly to Hackensack Water Company by calling our 24-hour emergency switchboard at 487-0011.

Thank you for your patience and cooperation while we work to improve your water system.

**Hackensack Water Company**

Figure 13-1 A sample letter notifying consumers of work to be done

**WATER SHUT-OFF NOTICE**

**ATTENTION OCCUPANTS**

**THE WATER SUPPLY TO THIS PROPERTY WILL BE SHUT OFF TEMPORARILY BETWEEN THE HOURS OF ________.M. AND________.M. ON ________ 20_____.**

THIS TEMPORARY SHUT-OFF IS NECESSARY IN CONJUNCTION WITH THE REHABILITATION OF THE WATER MAIN IN YOUR AREA. THIS REHABILITATION WORK IS A PLANNED CONSTRUCTION PROJECT AND IS DESIGNED TO IMPROVE EFFICIENCY OF THE WATER MAIN AND INFRASTRUCTURE. FOR FURTHER INFORMATION, PLEASE VISIT THE <WATER UTILITY> WEB SITE AT: ______________________________ .

IF YOU HAVE ANY QUESTIONS, PLEASE CALL THE <WATER UTILITY> CUSTOMER SERVICE DEPARTMENT AT____________________.

IF YOU NEED IMMEDIATE ASSISTANCE, PLEASE CALL ____________ (TITLE) AT (TELEPHONE#).

THANK YOU FOR YOUR COOPERATION.

<YOUR WATER UTILITY NAME HERE>

Figure 13-2 A sample water shut-off notice to be handed out to each customer

## COMMUNICATION NEEDS

Prior to a rehabilitation project, early planning of the customer notification program is an important priority. The simple objective of this communications program is to develop customers' understanding of how the proposed project will impact their water supply, the benefits associated with the work, the duration of the project including start date, and how the proposed project will impact their day-to-day activities.

The first step to maintain good customer/community relations is to determine who will be affected by the project. The project limits should be reviewed, paying special attention to actual shutoff or system isolation points. A rehabilitation project on one street may actually affect several streets or the entire neighborhood due to system pipeline layout or other operational issues. Secondly, an overall project schedule should be created so that advertising and public notices can be distributed before and during the actual work. This coordination requires careful attention to publishing schedules of daily and weekly newspapers as well as radio and TV station broadcast schedules. Last but not least, project notices should be sent out to the general public approximately three weeks before the construction starts and then on a weekly basis after that. This process will allow sufficient time for individuals and businesses to plan for the event and take measures to rearrange any special circumstances they may have and communicate with the utility directly prior to the construction. Program manager and construction contractor contact and project information can also be included on a utility's web site.

It is also important to review the affected customers' list. The utility should check to verify if there are any customers that need to be specially notified, such as hospitals, schools, nursing homes, public agencies, etc.

When the work to be done will impact historic areas, special business districts, or extremely busy subdivisions, a public meeting should be held to present the project. During this meeting the need for the project, its benefits as well as the overall process, can be explained. The meeting is a good place to share the time line (schedule), the name and contact information of the contractor, and other important information. This meeting will allow the impacted customers an opportunity to ask questions and possibly reduce the number of calls received due to misunderstandings about the work to be done. It also displays a willingness to work with the community and demonstrates the utility's desire to keep the distribution infrastructure current.

Most utility work causes some traffic disruption and the need for temporary detours. Detours should be identified and public notifications prepared for public announcements on newspaper, radio, television, web page, and other publications so that the traveling public, local rapid transit district, and emergency response agencies are aware of the work and can avoid the area. Also, the work area should be properly barricaded and caution notices posted (Figure 13-3). All OSHA safety rules for trench/pit excavations and traffic guidelines must be followed.

In addition to contacts with individual customers, overall community relations are equally important. Police, traffic control personnel, and transit district representatives must be included in planning meetings to discuss detours and work schedules. The fire department also must be made aware of the project and any effects on emergency access to fire hydrants or street closures. Hydrants removed from service should be covered or otherwise marked to indicate their unavailability prior to commencement of rehabilitation work.

## RESPONDING TO PROBLEMS

To ensure that potential construction problems are attended to in a timely manner, specific language should be included in the contract documents. Items such as customer courtesy, requirements for the response to problems, notifications to customers, communications, and site maintenance should be included.

It is best to have a dedicated field engineer or inspector on the job site at all times. In addition, for very large projects, it might be advantageous to dedicate an ombudsman or public relations person. This will ensure that any problems that do arise can be attended to as soon as possible. A phone number should be set up for the field engineer or inspector,

**CAUTION:**

**PLEASE BE CAREFUL WHILE DRIVING OR WALKING IN THIS AREA. WATER MAIN IS BEING CLEANED AND REHABILITATED FOR IMPROVED SERVICE.**

**Pipe aboveground is only temporary and will be removed as soon as possible.**

Figure 13-3 A sample caution notice to be posted at the work site

so that customers can call with questions or problems. Once a problem has been identified, both the field engineer or inspector, and the contractor should work together to address the issue. Individual responsibilities and response schedules should be identified so that sensitive issues do not go unresolved.

There should be a contact name for any nonwork-hour issues. This will allow the contractor to respond quickly to any leaks, service disruptions, or other service or project operational problems as soon as possible. This contact person should also be assigned for any emergencies that should arise at night.

As part of maintaining an effective attentiveness to all customer concerns, the contractor and the utility's inspector should maintain a customer response database listing all of the concerns that have been received as well as the date, time, and solution for the issue identified. This aids in reducing the number of calls that are not resolved.

## CONTRACT DOCUMENTS

### Construction Inspection/Quality Assurance

Inspection of the construction work is very important. The benefit of full-time construction inspection is that the field engineer or inspector ensures that the contract documents are adhered to and documents the progress of the work. This inspection work and documentation will keep the owner aware of the project milestones and progress.

The contract documents should clearly outline and define the material or product testing that is to be performed. This documentation will also specify the standard by which the product will be tested. The third-party testing firm should test and inspect materials, equipment, and work to eliminate anything that is defective. Documentation of the testing and/or inspection should be provided to the owner for compliance. If specific materials and equipment are not inspected or tested, a written agreement between the owner and contractor should be established or the nontested specific item should be removed. If requested by the owner, materials or equipment should be submitted for approval, acceptance, or inspection prior to their use on the project.

The contractor should keep the contract administrator/inspector informed by providing advanced notice of any upcoming material inspections or tests. If no owner-provided inspection is to be performed, the contractor must provide the progress of any material testing to the contract administrator. The contractor should furnish the contract administrator/inspector with proper authority for access to observe inspections and tests.

All test procedures should be reviewed by the contract administrator/inspector and confirmed with the contractor. All materials, equipment, and work should be inspected by the contract administrator/inspector or his/her designee. If any material, equipment, or portion of work does not meet the requirements of the contract documents, the material, equipment, or the portion of work, or any faulty portion, should be rejected and repaired/replaced at no expense to the owner.

## POST-CONSTRUCTION ACTIVITIES

### Video Inspection

Often, the pipeline to be rehabilitated (host pipe) should be inspected by a closed circuit television (CCTV) camera and video recorded. It may be advisable to have two inspections, one for the host pipe before the rehabilitation (pre-CCTV) and one after the pipe has been rehabilitated (post-CCTV). The CCTV equipment should be for potable water use only and certification of this should be provided to the owner's representative. The pre-CCTV inspection can be conducted before or after the cleaning process. This is dependent on the size, material, and condition of the host pipe. In most cases, pre-CCTV is completed after the host pipe has been cleaned to remove obstructions. Post-CCTV inspection of the rehabilitation work should be done after the work is performed and any service connections have been made but prior to returning the pipe to service.

The pre-CCTV equipment used for the inspection should have its own lighting system and the picture quality should be of quality to clearly show the walls of the pipeline. The video should have a distance indicator corresponding to the location of the equipment in the pipeline. The camera operator should take note of any unknown conditions, objects, or protrusions, which may damage the liner or make the installation difficult or impossible. A written log with these locations and a brief description of the obstruction should be kept. Any interference should be removed prior to rehabilitating the pipe. Discussion of potential rehabilitation work issues should be reported to the owner immediately so coordination and decisions can occur immediately.

The post-CCTV inspection should be done in one continuous video recording of the entire newly installed rehabilitation product. The camera operator should note any significant sags or defects in the rehabilitation. A written log of these locations and a brief description of the defect inside the new pipeline should be provided to the owner. In addition, a replacement process or suggested repair should be provided for discussion. If necessary, involvement with the manufacturer in the discussions will provide a complete analysis of causation of the defect and the proposed solution. Any major failures of the rehabilitated section of pipe (liner) should be addressed and a solution identified.

## Post-Construction Customer Satisfaction Survey

Another important aspect of the management of the work is to perform pre- and post-project inspections and data collection. This allows the flow readings and water quality parameters at each impacted customer to be documented before the work begins so that a baseline for the level of improvement can be presented when the work is complete. At the completion of the rehabilitation process when service to the customers is restored, the flow readings and water quality parameters should be taken again and documented as part of the final project report. Along with the improved service experienced by those impacted, this provides validation to the enhancement of the service delivered.

Job-completion customer surveys will provide an indication of how the contractor worked within the community and how well the project was completed from a utility customer standpoint. It can also be used to explain the benefits of the rehabilitation to the end user. This survey should be distributed to the affected parties (customers and agencies) following completion of the rehabilitation work. Typical rating categories could include: the adequacy of the notification of upcoming work, the pre-notification for any service shut-offs, perception of water quality before and after construction, courteousness of the contractor, effectiveness of communication with the impacted community, cleanliness of the job site, and overall performance should be included in the survey. A section for comments/suggestions for improvements should be provided. The rating survey should be divided into at least six categories to obtain adequate feedback from the community on how the work was performed for future projects.

## Contractor Performance Report Cards

Another method of determining the performance of the contractor is by rating the contractor's performance monthly. The contract administrator, the job site field engineer, or inspector could do the rating. The person to perform the rating should have experienced the most exposure and interface with the contractor to ensure the rating is accurate and not just an opinion developed as a snapshot of one point in time. This will provide the most fair and representative assessment. Typical areas of rating could include, adequacy in providing timely shut-off notices to the customers; communication with other impacted agencies; responsiveness to problems in the field; effectiveness in planning and management of the rehabilitation work; success in providing a safe construction site for the public as well as the staff performing the work; maintaining the job site in a clean and orderly state; job site restoration; and participation in resolution of any cost disputes. A letter or numerical rating could be assigned to each category.

## Measuring/Demonstrating Program Benefits

The benefits of pipeline rehabilitation include enhanced water quality by replacing rough, corroded inner surfaces with a smooth lining (coating). Another common benefit is the elimination of discolored water due to internal pipe corrosion as well as the reduction of an environment that could encourage and allow bacteria to grow. The rehabilitation of the pipeline also increases flow characteristics for fire flow and in certain methods and materials may extend the life of the pipelines for up to 50 years. A rehabilitated pipeline can dramatically reduce distribution system energy requirements. Pipeline rehabilitation can also provide a potential cost savings compared to conventional pipeline replacement depending on in-situ conditions. Depending on the rehabilitation technique selected and the number of service connections, rehabilitation can be completed in less time with less disruption to customers. Effective program management and communication is the key to the success of any project. Utilizing the techniques previously listed to strengthen communication between the customers, contractor, and owner before the project, at the beginning of the rehabilitation work, and throughout the process will lay the foundation for a successful rehabilitation project.

When properly researched, selected, and utilized, the rehabilitation options identified in this manual will enhance a utility's ability to provide effective service to their customers. There are no "one approach fits all" responses in water main rehabilitation. For this reason, water industry professionals have developed various processes to assist in the utility's challenge of maintaining a sufficient supply of potable water at an adequate pressure to the customer in a cost-effective manner.

This page intentionally blank.

# Appendix A

# Structural Lining Design Issues

## STRUCTURAL CLASSIFICATION OF LINING TECHNIQUES

Lining systems used to rehabilitate potable water pipelines can be classified into four groups according to their effect on the performance of the lined pipe when subjected to internal pressure loads.

### Class I Linings

Class I linings are essentially nonstructural systems used primarily to protect the inner surface of the host pipe from corrosion. They have no effect on the structural performance of the host pipe and have a minimal ability to bridge any existing discontinuities, such as corrosion holes or joint gaps. Hence, they have minimal effect on leakage. Their use is indicated in pipes suffering from internal corrosion or tuberculation but still in structurally sound condition and where abatement of current or future leakage is not an issue. Examples of Class I linings are cement–mortar lining and epoxy resin lining.

### Class II and III Linings

Class II and III linings are both interactive and semistructural systems. When installed, the liners closely fit the host pipes, and any remaining annulus is rapidly eliminated when internal operating pressure expands the lining. Because the stiffness of such a lining is less than that of the host pipe, all internal pressure loads are transferred to the host pipe, leading to their classification as interactive. Such a lining is required only to independently sustain internal pressure loads at discontinuities in the host pipe, such as corrosion holes or joint gaps, or if the host pipe is subject to structural failure.

A liner is considered to be in Class II or III if its long-term (50-year) internal burst strength, when tested independently from the host pipe, is less than the maximum allowable operating pressure (MAOP) of the pipeline to be rehabilitated. Such a liner would not be expected to survive a burst failure of the host pipe, so it cannot be considered as a replacement pipe. However, Class II and III liners are designed to bridge holes and gaps in the host pipe on a long-term basis, and various systems can be further classified in terms of the magnitude of the holes and gaps they can bridge at a given MAOP. Some systems are capable of bridging holes and gaps of up to 6-in. (152-mm) on a long-term basis at an MAOP of 150 psi (1,034 kPa).

The separation of these systems for spanning holes and gaps into two classes is based on their inherent resistance to external buckling forces and dependence on adhesion to the host pipe wall. Class II systems have minimal inherent ring stiffness and depend entirely on adhesion to the pipe wall to prevent collapse if the pipe is depressurized. Class III liners have sufficient inherent ring stiffness to be at least self-supporting when depressurized without dependence on adhesion to the pipe wall.

Class III liners can also be designed to resist specified external hydrostatic or vacuum loads.

Use of Class II or III linings may be indicated where the host pipe is suffering from one or more of the following conditions:

1. Severe internal corrosion leading to pinholes and leakage
2. Leakage from faulty joints
3. Localized external corrosion resulting in pinholes and leakage

Although the liner will not prevent further external corrosion, it will prevent leakage at corrosion holes. This capability guards against the associated effects of that leakage on the exterior of the pipe and the corrosivity and support offered by the surrounding soil.

### Class IV Linings

Class IV linings, termed fully structural or structurally independent, possess the following characteristics:

1. A long-term (50-year) internal burst strength, when tested independently from the host pipe, equal to or greater than the MAOP of the pipe to be rehabilitated
2. The ability to survive any dynamic loading or other short-term effects associated with sudden failure of the host pipe due to internal pressure loads

Class IV linings are sometimes considered to be equivalent to replacement pipe, although such linings may not be designed to meet the same requirements for external buckling or longitudinal/bending strength as the original pipe. Also, they may be of smaller internal diameters. Class IV linings can, of course, be used in circumstances similar to those for Class II and III, but their use is essential for host pipes suffering from generalized external corrosion where the mode of failure has been, or is likely to be, catastrophic longitudinal cracking.

Some available renovation technologies can offer both Class II and III and Class IV linings, while a given lining system may be rated as Class IV for MAOP levels up to a threshold value and Class II and III for higher pressures.

## Additional Design Considerations

In addition to internal pressure loads, linings may also be required to sustain external buckling loads during periods when the host pipe is depressurized, as well as transient vacuum loads. Some systems (Classes III and IV) can be designed to offer significant inherent resistance to such external loads, while others (Class II) depend solely on adhesion to the host pipe wall. Inherent resistance to external buckling normally varies with increased lining thickness and hence cost. Care should therefore be taken to ensure that such performance requirements are accurately defined.

The hydraulic performance of the lined pipe will be determined by the thickness of the liner, its closeness of fit to the host pipe, and its internal smoothness (*C* value). The lining process is usually preceded by extensive cleaning, which will itself restore the original flow cross section of the pipe. Liners of plastic materials are significantly smoother than the inner surface of a deteriorated host pipe, and they may even be smoother than the original pipe. In addition, many lining systems provide essentially joint-free coverage over long sections, so they offer fewer disturbances to flow than jointed sections of pipe. In general, close-fit plastic lining systems with a standard dimension ratio of 26 or more normally retain the original flow capacity of the pipe.

This page intentionally blank.

# Index

NOTE: *f.* indicates figure; *t.* indicates table.

# AWWA Manuals

M1, *Principles of Water Rates, Fees, and Charges*, #30001
M2, *Instrumentation and Control*, #30002
M3, *Safety Management for Water Utilities*, #30003
M4, *Water Fluoridation Principles and Practices*, #30004
M5, *Water Utility Management*, #30005
M6, *Water Meters—Selection, Installation, Testing, and Maintenance*, #30006
M7, *Problem Organisms in Water: Identification and Treatment*, #30007
M9, *Concrete Pressure Pipe*, #30009
M11, *Steel Pipe—A Guide for Design and Installation*, #30011
M12, *Simplified Procedures for Water Examination*, #30012
M14, *Backflow Prevention and Cross-Connection Control*, #30014
M17, *Installation, Field Testing, and Maintenance of Fire Hydrants*, #30017
M19, *Emergency Planning for Water Utilities*, #30019
M20, *Water Chlorination/Chloramination Practices and Principles*, #30020
M21, *Groundwater*, #30021
M22, *Sizing Water Service Lines and Meters*, #30022
M23, *PVC Pipe—Design and Installation*, #30023
M24, *Dual Water Systems*, #30024
M25, *Flexible-Membrane Covers and Linings for Potable-Water Reservoirs*, #30025
M27, *External Corrosion: Control for Infrastructure Sustainability*, #30027
M28, *Rehabilitation of Water Mains*, #30028
M29, *Water Utility Capital Financing*, #30029
M30, *Precoat Filtration*, #30030
M31, *Distribution System Requirements for Fire Protection*, #30031
M32, *Computer Modeling of Water Distribution Systems*, #30032
M33, *Flowmeters in Water Supply*, #30033
M36, *Water Audits and Loss Control Programs*, #30036
M37, *Operational Control of Coagulation and Filtration Processes*, #30037
M38, *Electrodialysis and Electrodialysis Reversal*, #30038
M41, *Ductile-Iron Pipe and Fittings*, #30041
M42, *Steel Water-Storage Tanks*, #30042
M44, *Distribution Valves: Selection, Installation, Field Testing, and Maintenance*, #30044
M45, *Fiberglass Pipe Design*, #30045
M46, *Reverse Osmosis and Nanofiltration*, #30046
M47, *Capital Project Delivery*, #30047
M48, *Waterborne Pathogens*, #30048
M49, *Butterfly Valves: Torque, Head Loss, and Cavitation Analysis*, #30049
M50, *Water Resources Planning*, #30050
M51, *Air-Release, Air/Vacuum, and Combination Air Valves*, #30051
M52, *Water Conservation Programs—A Planning Manual*, #30052
M53, *Microfiltration and Ultrafiltration Membranes for Drinking Water*, #30053
M54, *Developing Rates for Small Systems*, #30054
M55, *PE Pipe—Design and Installation*, #30055
M56, *Nitrification Prevention and Control in Drinking Water*, #30056
M57, *Algae: Source to Treatment*, #30057
M58, *Internal Corrosion Control in Water Distribution Systems*, #30058
M60, *Drought Preparedness and Response*, #30060
M61, *Desalination of Seawater*, #30061

This page intentionally blank.

www.ingramcontent.com/pod-product-compliance
Lightning Source LLC
LaVergne TN
LVHW082350200926
848481LV00012B/3515
* 9 7 8 1 5 8 3 2 1 9 7 0 6 *